Work, Ritual, Biography

Work, Ritual, Biography
A Muslim Community in North India

DEEPAK MEHTA

DELHI
OXFORD UNIVERSITY PRESS
CALCUTTA   CHENNAI   MUMBAI
1997

Oxford University Press, Walton Street, Oxford OX2 6DP
Oxford New York
Athens Auckland Bangkok Calcutta
Cape Town Chennai Dar es Salaam Delhi
Florence Hong Kong Istanbul Karachi
Kuala Lumpur Madrid Melbourne
Mexico City Mumbai Nairobi Paris
Singapore Taipei Tokyo Toronto
and associates in
Berlin Ibadan

© Oxford University Press 1997

ISBN 0 19 564021 7

Typeset by Rastrixi, New Delhi 110070
Printed in India at Rekha Printers Pvt. Ltd., New Delhi 110020
and published by Manzar Khan, Oxford University Press
YMCA Library Building, Jai Singh Road, New Delhi 110001

In memory of
Sri Krishna Mehta (1910–1986)

Acknowledgements

This monograph, an outcome of my doctoral field research, owes much to individuals and institutions who have been more than generous with their support, at once intellectual, emotional and financial. Part of my research was funded by a Junior Research Fellowship and part by friends and colleagues.

The field trip to Barabanki would not have been possible without the help of Masood Khan and his natal and conjugal family. Muhammad Mushir Khan, in particular, provided me with a second home. It is difficult for me to express in words my gratitude to his family. Azeer and Maudood accompanied me on my initial reconnaissance trips and plugged vital gaps. Sadiq Ali began my education in working on the loom while Haji Ghulam Nabi showed me how the tradition of weaving is constituted. Miriam and Muhammad Umar nursed me through an illness and their son, Imtiaz, attempted to complete what Sadiq Ali had begun. To Sufi Baba I owe a special debt of gratitude and friendship. At Lucknow Yash provided welcome relief from the rigours of fieldwork and Ajay and Uma, convinced I was malnourished, cooked up elaborate meals at short notice. To all of them my *salam*.

The Friday Colloquium of the Department of Sociology, Delhi University, where most of this monograph was presented, proved a rich source of critical comment. The discussions, at times gladiatorial, were mostly stimulating and always intimidating. I thank the participants for their views.

Jit Uberoi's critical insights and elliptical suggestions have supplied the foundations of the present work. Veena Das has been instrumental in moving this research along pathways that were novel to me. The initial enthusiasm of the two of them has never flagged.

Shahid Amin, Gyan Pandey, Amitava Ghosh, Rabindra Ray and Amrit Srinivasan have in their separate, and no doubt idiosyncratic, ways helped in the resolution and indeed complication of various problems. Neil May, Rita Brara, Sanjay Nigam, Felix Padel, and Anant Giri all read parts of this monograph and pointed out problems, both intellectual and stylistic.

Savyasaachi, Punam Zutshi and Bhatia *sab* have all read more than one draft of this monograph. Khalid Tyabji, Roma Chatterji and Rajendra Pradhan, hard-pressed for time themselves, provided support and friendship during a dilatory period. Krishnaswamy, Harish Naraindas and Sangeeta Chattoo have patiently heard me out on numerous occasions. Much of my conversations with Sadhana Bery and Paola Bachhetta were conducted in absentia.

To Jaya and Tushar, a special bond of friendship and love.

Contents

	Note on Transliteration	xi
1.	The Ansaris of Barabanki	1
2.	Household, Kin, Work and Ritual in Barabanki	36
3.	The Semiotics of Weaving	74
4.	Work, Worship, Word: A Study of the Loom	115
5.	Women's Work: Quilt Making and Gift Giving	143
6.	Circumcision, Body and Community	178
7.	Inner Voice and Outer Speech: The Life History of Sufi Baba	214
8.	Conclusion	241
	Glossary	261
	References Cited	273
	Index	279

Note on Transliteration

Hindustani terms are used in their English version. When used the first time they have been italicized. Hindustani terms are formally transliterated in the appendix. The system of transliteration is based on Platts *A Dictionary of Urdu, Classical Hindi and English*, Oxford University Press 1884. The way such terms are spelt in the dictionary is found in the Glossary.

Chapter One

The Ansaris of Barabanki

THE PROBLEM

This book studies a community of Muslim handloom weavers living in the district of Barabanki, UP. The study specifically focuses on the artisans of two villages of the district, Mawai and Wajidpur. Situated approximately seventy-five kilometres east of Lucknow, the Muslim section of these villages is composed mainly of weaving families. They use the eponym 'Ansari', though they are better known, both in the literature and by those living alongside them, as 'Julaha', a term that designates their low-caste status and associates it with a series of pejorative stereotypes: 'dim witted', 'sectarian', 'easily provoked to violence' (Crooke 1974; see also Pandey 1990 for a critique). Absent in this designation is the Ansari conception of work, their claim to an Islamic heritage and their representations of themselves as men and women and so on. In addressing such issues this study aims to provide a better understanding of the Ansari community, though for obvious reasons the picture presented is partial, perhaps idiosyncratic, and one which could be drawn in ways other than those attempted here.

The data for this study was collected in the course of field work (1985–6) in Barabanki. I selected a community of Muslim artisans who manufacture handloom and *khadi*[1] cloth as part of their traditional craft of weaving. This selection was determined by two considerations. First, I was interested in identifying a community whose means of livelihood was also self-consciously a way of life. Second, if this was the case, I wanted to explore a series of relationships: between the group's system of production and its social organization, between instrument and group, between language and work. These relationships presumed a larger question:

---

[1] *Khadi* is cloth that is both handspun and handwoven.

2 • *Work, Ritual, Biography*

what is a mode of 'doing' and how is it possible to articulate such non-discursive practices?

Once field work began, it became evident that the life of the Ansari weavers could not be formulated by reference to a typology of action which separated work from non-work, the physical from the verbal gestures of weaving, and indeed the instrumental from the expressive aspects of the craft. This lack of sharp boundaries was apparent in at least two significant rituals: circumcision, and weaving cloth for the shroud. While the first is indirectly connected with weaving, the second is of crucial importance in understanding the cosmology of weaving. In effect, an exploration of this question — what is a mode of doing? — implied an examination of the world fashioned by weaving, and how factors external to weaving, such as rituals and prayer and other forms of work, fed into it. This is not to suggest that weaving denotes the paradigmatic identity of the Ansari community (though it may be seen thus), but to show that the relationship between the working and non-working life of the weaver has a consistency and coherence which can be argued ethnographically.

In the present ethnography, each chapter deals with themes which recur in other chapters. These will be discussed later. Theoretically, I will argue that in their mode of work, as much as in their ritual and festive life, the Ansaris combine 'discourse' and 'practice' in a way that is not disjunctive. In their practice they are like Bourdieu's Algerian peasants, but in their discourse they are a literate people who belong to a great discursive tradition.[2]

As used here, practice refers to non-verbal gestures and techniques inherent in the relationship between man and machine. Such practice identifies the instruments of work, including the body and the operation of these instruments. Work refers to two areas of productive labour within the household: cloth production and quilt stitching. Outside the domain of production, practice refers to a mode of behaviour through which actors engage with

---

[2] Bourdieu (1977) presents his theory of practice on the basis of fieldwork among the Kabyle of Algeria. In this theory, the Kabyle Algerians are important in as much as they show a way of doing things, not because they are Muslims. The present work tries to combine 'doing' with the issue of being Muslim. What it means to be Muslim is elicited mainly from conversations, but in a few crucial cases from observation.

the social world in a non-discursive way. Broadly, this engagement situates the actor's body within the frame of accepted norms. This behaviour highlights the arena of 'doing'. The domain of discourse identifies gestures that are verbalized or have the potential of being 'said'. In the production of cloth and quilts such discourses make available an Islamic tradition of worship and prayer. But they also establish a dialogic relation between the addressers and addressees of such prayers. In both cases, discourse is an inherent facet of work and is marked on the body of workers. Outside such work, discourse shows how the community's presence is mapped onto its individual members: as in the case of work, it is embodied. Such inscription is, however, problematized, as we will see in chapter seven.

The domain of doing may be studied under the general rubric of practice, and that of saying under the term discourse. The latter valorizes language and studies non-linguistic systems as if they were languages,[3] but it does not provide an adequate account of non-verbal action from a perspective internal to that action. Thus, action is studied as part of a general theory of language. For this reason, non-verbal action conforms to a logic available in linguistics. Rather than argue that action is amenable to a semiotic operation in the Saussurian sense, I will show that there is a complex interrelation between action as technique and the positing of this technique.

The technique of making cloth and quilts is constituted within a semiotic system. That is to say, such techniques can be studied as comprising a system of signification. In turn, signification can be delimited in the way that Saussure's theory of 'langue' shows. However, it is important to bear in mind that techniques also constitute their own systems of meaning. In this sense, action would comprise of a verbal and a non-verbal part. Further, it may be convenient to show that non-verbal action is embodied while verbal action is disembodied. Rather than arguing for a rift between the verbal and the non-verbal, I aim to show that the site on which they come together is the body of individual members of the community and of the collective taken as a whole. In effect, the body is not merely a legible picture translated into that which

[3] Here I wish to argue against those types of approaches that employ the Saussurian model of language to understand non-linguistic phenomena. See, for instance, Munn (1973), Humphrey (1971).

4 • Work, Ritual, Biography

can be said in language, but also an object that produces its speech. In this sense I have asked the question: What is a way of making? An investigation of this question shows that inherent in non-verbal practice is a mode of action which combines the technique of the worker with the instruments of work. Outside work, such practice shows how the discursive codes of the community acquire an instrumentality in relation to such actions. The relation between the technique of making or doing and the coding of such techniques as techniques is simultaneously a discourse (in that the written and spoken word inheres in various physical gestures) and practice (in that the fabrication of material objects in work and bodies outside it obey a prescribed way of doing).

METHOD AND APPROACH

To grasp the formal structure of the mode of weaving cloth and stitching quilts and their relationship with the non-working life of the Ansaris, I will carry out three sorts of investigation. The first shows how work and ritual in the domestic domain is constituted by describing the social structure of the community. Accordingly, chapter two examines the relationship of work and ritual to the kinship institutions of the Ansaris. The second investigation shows the logic of various practices from a perspective internal to them. The next three chapters examine the world constituted by weavers and weaving (chapters three and four), and by quilt makers and quilt stitching (chapter five). The third sort of investigation stands outside the domain of work and situates the community in its ritual life, on the one hand, and through the eyes of one individual, on the other. Chapter six discusses the ritual of circumcision and the place of the community within it, while the next chapter attempts a partial biography of one Ansari individual, Sufi Baba, in his relationship to the community. The ritual reproduces the social structure of the community, but in the case of the individual, the boundaries of this structure become porous. Both these chapters focus on one case study each, and attempt to present an ethnography of individuals. In this sense, the ethnography of the study moves from a general account to particular descriptions.

I use 'discourse' in the way Bakhtin (Bakhtin/Medvedev 1985;

Volosinov 1973) has developed it, and 'practice' in Bourdieu's (1977) conception. In his writings on language Bakhtin seeks to transcend the dichotomy arising from 'individualistic subjectivism' and 'abstract objectivism' (Volosinov 1973: 48), just as Bourdieu, in elaborating a theory of practice, cautions both against the overdetermination of the rule and a 'naive subjectivism' (Bourdieu 1977: 3–4). For Bakhtin, abstract objectivism rejects the 'speech act' or the utterance as an entirely individual phenomenon by privileging the formal (i.e. the syntactic and rule-bound) character of language, while individualistic subjectivism valorizes it precisely for this reason and attempts to explain the utterance in terms of the psychic life of the speaker (Volosinov 1973: 82). In place of this dichotomy Bakhtin proposes the doctrine of 'finalization' which combines a study of discourse in its compositional and thematic details. Taking the utterance, not the sentence, as his point of departure, he argues that the word is always oriented towards the addressee and for this reason includes 'dialogue' (Volosinov 1973: 85). In considering 'practical knowledge' as his point of departure, Bourdieu seeks to bridge the 'ritual either/or choice between objectivism and subjectivism' (1977: 4) by proposing lasting dispositions (habitus) to action, on the one hand, and the place of temporality and improvization, on the other. Instead of emphasizing the rule-bound character of social life, he thinks it more appropriate to understand the generative schemes that lead to the strategies by which practices arise, or, in his words, how the 'orchestrated improvization of common dispositions' (1977: 17) reproduce the divisions of the social structure on an everyday basis.

Practice

In building a theory of practice, Bourdieu distinguishes between 'habitus' and 'objective structures'. The habitus is a 'system of lasting, transposable dispositions' that functions as a 'matrix of perceptions, appreciations and actions' (1977: 82–3). The objective structure is the frame within which the habitus is realized. The habitus generates homologous formations across different social activities, such as funerals, ploughing, harvesting, circumcision, marriage, etc. Through the habitus, different conceptual

6 • Work, Ritual, Biography

fields are organized by the same set of symbolic relations — the cooking calendars, the farming calendar, the daily cycle, the life cycle, etc. (ibid: 143–6). The symbolic relations and modes of practice organizing these different fields are not only schematically equivalent, but also variants of a single type of structure (ibid: fig 9). The importance of the habitus is that it is logically prior to actual events of practice and is simultaneously subject to strategic manipulation in practice. For this reason it is both product and resource.

The habitus, thus, generates practices. Practices, in turn, reproduce specific objective conditions that lead to their generation via the habitus. Practices cannot be deduced directly either from objective conditions, or from conditions establishing the lasting principle of their existence. Practice, as inscribed in time, is always emergent and unfinished (ibid: 9) because formal, objective rules yield to strategies. In turn, strategies transmute individual interests into collective ones. Bourdieu gives a distinctive meaning to the term strategy. By this he does not point to choice, intention or calculation. There is no prediction, but only an 'assumed world' as the repetition of the past. Strategies arise because actors do not always know what they are doing, and often what they do has more meaning than what they say (ibid: 79). For this reason, too, knowledge of practices lies in practices themselves. These practices are dominated by the acquisition and maximization of material and symbolic capital, on the one hand, and the development of the body, both individual and collective, on the other. The first is the basis of patrimony while the second leads to duration (through its fertility) and space (through its activity). The proliferation of strategies is related to this politics which seeks to reproduce, through wealth and body, land and heirs. Through this politics the individual is annexed to the group.

At the heart of this transmutation are temporal rhythms and forms which structure both the group and its representation of the world. Temporal rhythms fulfil the function of integration by instituting divisions and by giving to practice the appearance of 'realised myth' (1977: 163). By a realised myth is meant that the belief of the group is grounded in a practice which reconciles subjective demand with collective necessity. Variations in prac tices can be understood through 'practical taxonomies', which are a transmuted form of the real divisions of the social order, and which contribute to its reproduction through 'orchestrated practices'. The latter reproduce the social world on an everyday basis by making it appear mundane and natural, as much as they legitimate 'natural' divisions — of gender, age, generation, etc. — of the social order by socializing individuals into them.

Bourdieu's main concern, then, is to explain the necessary relation of practices to objective structures. This is achieved by looking at the genesis of practices and rests on the acquisition of knowledge. In practice this genesis implies an internalizing of structures and an externalizing of the habitus. A temporal dimension is thus introduced: practices expressing the habitus correspond adequately to situations exemplifying the objective structure only if the latter is stable during the process of internalizing/ externalizing. If not, practice resembles the structure at the preceding point, the one at which it was internalized by the habitus. However, these objective structures can be understood only through the discourse of the sociologist, a problem that Bourdieu does not address. While structures change the habitus cannot. Because structures are not given to actors the traditional image of peasant societies is reinforced — the only history written on them is of an alien order.

In mapping how practices originate Bourdieu shows a way of talking of non-discursive practices. In so far as the habitus generates practice I will show the dispositions that constitute them. These dispositions are found most clearly in the constitution of the worker's body in the act of weaving, as well as the novice's body in the circumcision ritual. Further, on every occasion of making cloth and quilts practical taxonomies inscribe on the body the divisions of the social structure. We find, however, a second inscriptional process, one that both transcends the normative limits of the social structure and interrogates it. Some of the discourse on quilt making questions the divisions effected by weaving. This is still more marked in the biography of Sufi Baba whose discourse critiques the world constituted by weaving.

In establishing a theory of practice, Bourdieu propounds a knowledge of practices that its actors do not reflect. They witness it without appropriating it. Such knowledge of practice as does exist is known only by people other than its bearers. However,

8 • *Work, Ritual, Biography*

Bourdieu's insight that practices are always unfinished and emergent is crucial in understanding the productive and reproductive character of non-discursive practices. It is crucial not only because a discourse is constructed after the event, but because such practices themselves produce a discourse. Unlike Bourdieu's peasant, however, the discourse of the Ansari weaver in relation to weaving, self-consciously proclaims its Islamic heritage and derives its credibility from the *Mufidul Mu'minin*, the sacred book of the weavers. This book acquires a voice in the practices of weaving, when the prayers mentioned in it are uttered by weavers. In this sense words are linked to physical gestures. Together, words and physical gestures reveal the domain of action in work.

Using Bourdieu's theory of practice as a point of departure it is possible to ask how words and physical gestures come together. For this, it is important to recognize that a knowledge of the various practices of weaving is not the sole prerogative of the sociologist but the product of an interaction between the sociologist and his subject of study. It also questions the notion of a whole whose unitary nature is axiomatic in formalist approaches. It involves a recognition of the other, or the addressee, as part of the same discourse. One way of constructing a discourse about practice is by considering Bakhtin's theory of utterance, which unfolds through an examination of its internal coherence and its location within wider society. The internal coherence of the utterance is given in the theory of finalization, while its relationship to its period is found in the idea of dialogue.

DISCOURSE

Finalization is the process by which a discourse becomes complete, but it should not be confused with ending. Finalization is not positionally linked to the end of the discourse, but provides its frame and sets it off from its context as it unfolds (Bakhtin/Medvedev 1985: 130). Discourse is finalized through 'theme' and 'composition', linked by the mode of discoursing. Thematic finalization corresponds to 'story' completion, and the latter to 'plot' completion. Story is the sequence of events or referents comprising the thing described and plot is the structuring of reference within a narrative. This means that the theme is of the whole utterance as a definite socio-historical act and cannot be studied on the same plane as phonetics and syntax. For this reason, thematic unity is inseparable from circumstances of place and time (Bakhtin/Medvedev 1985: 132). Compositional unity is the syntactic structuring of the work. In contrast to the theme it points to all the moments of the utterance 'that are reiterative and identical unto themselves in all their repetitions' (Volosinov 1973: 120). Signification has the potentiality of signifying in a concrete theme, but in itself signifies nothing.

Every utterance, then, has two aspects: that coming from language is reiterative, and that, deriving from circumstance of place and time, is unique. The latter distinguishes the utterance from the sentence. Unlike the sentence, it is related to a speaker, an object and crucially, to previous and future utterances with which it dialogues. The constituent elements of discourse, then, would be the utterance, the speaker and the listener.

The term Bakhtin uses to designate the relation of every utterance to other utterances is dialogical. Moving from this elementary level he focuses on 'reported speech' as 'speech within speech, utterance within utterance, and at the same time also speech about speech, utterance about utterance' (Volosinov 1973: 115). To explain this definition he explains the difference between 'linear' and 'pictorial' styles of speech (ibid 1973: 120). The former constructs clear-cut contours for reported speech, while the pictorial style obliterates these contours and is individualized to a far higher degree. Within this style one of the voices can be dominant.

The placing of the dialogical utterance in a temporal span referring to the past, present and future implies that the thematic unity of the text can never be final. Thus, the dialogical context knows no limits. Yet Bakhtin does not advocate a limitless relativity. While every utterance has a receiver who is a 'second', it also has a 'third', a 'super-receiver', who, depending upon the historical period, may be God, the absolute truth, history, etc. Every dialogue, then, occurs against the backdrop of a present but invisible third entity. For this reason, the 'third' is the constitutive moment of the whole utterance and proceeds from the nature of discourse that is always in search of response. In this sense the other plays a decisive role.

Bakhtin's theory of the utterance and his emphasis on the

necessity of dialogue are important analytical devices used in this text. Through the concept of finalization I argue for the coherence and attempted completeness of discourses generated both by various practices and on them. Through the concept of dialogue I show that these discourses are open-ended, reiterative and sometimes antagonistic. Of crucial importance to dialogue is the presence of reported speech. Its significance for the discourse of work, the circumcision ritual and the biography is that it organizes the stylistic variants of the utterance and is an objective document of the social reception of the utterance.

This is most apparent in the case of the *Mufidul Mu'minin*, a text documenting the origin and consolidation of the sacred work of weaving. It is used by the weavers of Barabanki on different occasions. In describing how this text is used we learn of the 'steadfast social tendencies in an active reception of the other speaker's speech' (Volosinov 1986: 117). These tendencies are evident in the overt dialogue between interactants and within the same speaker. In both cases, reception invests the discourse with dialogue in Bakhtin's use of the term. We find a similarity between active reception and Bourdieu's description of habitus. The text is a kind of habitus which generates the various practices of weaving. But it acquires an instrumentality through the practices of weaving. Such instrumentality establishes a multifaceted dialogue: between the addresser and addressee, between the individual and the community (most evident in the case of Sufi Baba) and between the community and the patron saint.

Both discourse and practice are unfinished and emergent. For Bourdieu the rule is substituted by strategy. Strategies cannot be fixed in a unitary structure, characterized as they are by improvisation. Similarly, Bakhtin claims that the utterance cannot be fixed solely in an objective syntactic structure, but achieves fruition through the social evaluation of discourse. This is seen most clearly in the Ansari conception of work. Work is not only composed of a heterogeneity of meanings, but is also of an emergent nature. This does not lead to an incapacity to generalize. It points to the argument that the same work connotes differently in different contexts. Consequently both weaving and quilt making are anchored in plural modes of organization.

Just as both weaving and quilt making are placed within their setting as much as they reproduce their setting, so also the ritual of circumcision shows how the community marks its presence on its individual members and is simultaneously validated through the domestic group. We find plural interpretations of the mark of the community in that men provide one account (primarily verbal) of the ritual while women provide another account that is located in a gestural and embodied medium. The case of Sufi Baba is more complex. As healer of various maladies he creates a context within which the body in pain can be situated. But as a significant participant in the most important festival of the Ansaris he becomes the subject of signs — the framing procedures adopted by the community *vis-à-vis* him seek to deny him agency.

THE CONCEPT OF THE EVERYDAY

Before I discuss the range of meanings available in various practices and the discourses around them it may be stated that weaving and quilt making, on the one hand, and the ritual of circumcision and the biography of Sufi Baba, on the other, unfold in two domains: the everyday and extraordinary world of the Ansaris.[4] My discussion of the 'everyday' is based on Heller's (1984) use of the term. Heller defines everyday life as the aggregate of individual reproduction factors which become the impetus to social reproduction (Heller 1984: 4). This points to the reproduction of the person and his group. The reproduction of the person is accomplished by his growing into an already fashioned world and by transmitting knowledge of this world to others. In this way the person objectivises himself through spontaneous and generalized activities. Objectification means that the person externalizes his capabilities which then detach from their human source. Participation in generalized activities is the basis of transcending the everyday. All objectification which does not relate either to the person or to his immediate environment exceeds the boundary of the everyday.

A second characteristic feature of everyday life is its heterogeneity: the norms and rules for action contingent on the

---

[4] I use the term everyday and extraordinary deliberately, rather than the more convenient sacred and profane, or ritual and mundane, in an effort to show that the everyday incorporates in its operation ritual and sacred elements.

individual are varied and dispersed. This does not, however, exclude homogeneous zones for action. In fact, the more generalized an objectification the more homogeneous it becomes and the more generic the act the greater the absorption of the person in a single sphere of action (Heller 1984: 56–7). Among the five zones Heller identifies as generic, work provides the first example.[5]

In Heller's estimation, work refers to both an everyday and generic activity. While the act of work (in the sense of a technique of doing and making) situates it as an organic part of everyday life, work as activity is generalized since it is common to the human species. As an everyday act work is labour. Through labour, done on a regular basis in a given time, the worker supports himself. This notion of work involves an attitude of 'wear and tear', of both the body and the intellect. To the extent that everyday life is the reproduction of the person and the group human beings must work. For this reason work is the process of producing a product which can satisfy the needs of others through material production. It is also guided by objectification since in making the product the worker is limited by a homogenizing and generalized activity.[6]

This is the broad frame within which the category of work for the Ansaris may be placed. It refers to two types of acts that are oriented towards the production of cloth and quilts: those that are located in the everyday world of the Ansaris and are heterogeneous and those which, emerging from the everyday, transcend it. These activities are homogeneous. A significant omission in

[5] The other four are: morals; religion, politics and law; science and philosophy; the arts.

[6] In an earlier article, Heller (1981: 71–9) echoes Arendt's (1958) distinction between work and labour. Arendt distinguishes between specialization of work — the pooling of individual skills in making a finished product — and division of labour, which quantifies the energies of individual workers who in practice are interchangeable. In modern society, Arendt argues the two are separate though they are confused as identical. The two, for Arendt, can only be united when workers live and act together. Work, as Arendt understands it, is a seminal act by which humans distinguish themselves from nature and assert mastery over it. On the other hand, quoting Marx, she says that labour is 'man's metabolism with nature' (Arendt 1958: 98) — an act that arises out of the biological necessity of staying alive and reproducing the species.

this conception is the relationship of the worker to the instruments and objects of production. In this study, this relationship is delineated through the body techniques of weaving. Further, the relationship of the material object with the community which uses it is introduced by referring to the material culture of work. Finally, the place of tradition or custom in determining work in everyday life is apprehended through the tradition of weaving.[7]

Just as work is situated in the everyday and extraordinary world of the Ansaris, so also the ritual of circumcision and the everyday discourse surrounding the fact of being circumcised are similarly located. The discourse on circumcision in everyday life is characterized by the emergence of a 'master narrative', one which separates Muslims from non-Muslims and self-consciously locates the Ansari community within a particular understanding of Islam. The ritual both opens out and limits the everyday discourse on circumcision. The opening out is evident in that the ritual admits of multiple interpretations, all of which refer to the ritual wound. The closure is seen in that a referential object is strongly marked in the ritual, an object that is lost in the everyday discourse. In this sense, the everyday life of the Ansaris, in relation to their extraordinary world, is characterized by a relative absence of referentiality. Here, the 'homogeneous' (or what I call the extraordinary) does not emerge from the everyday, but the two are simultaneously posited. It is almost as if a metonymic link exists between the ritual and the everyday discourse surrounding it, but a link where the everyday selectively incorporates the ritual so that the latter is invested with an incorporeal and discursive valuation.

The everyday/extraordinary distinction is also played out in Sufi Baba's relationship to the Ansari community. As healer, he

---

[7] Heller mentions that work, language and custom share a common characteristic: every utterance, every customary act, every 'job of work done' is a generalization (1984: 66). Here, language, custom and work are three distinct and homogeneous spheres of action. Contrarily, I argue that language and custom are inherent in the act of work. This inherence is examined by seeing how work is constituted as both discourse and practice. To discuss the relationship of work to discourse I will focus on the three domains outlined above: the tradition of work, the place of the material object in the life of the community, and the relationship of the worker to the instruments and objects of work.

interacts on an everyday basis with the women of the community, while his public and more dramatic persona is visible in an annual festival where he carries out an elaborate performance as clown and spokesman of the community, and later as mediator between customer and prostitute. In his life the relation between the everyday and the extraordinary is played out almost serially in the relation between the public and the private, between his penchant for authorizing signs and being authored by them, and in his mode of worship, which oscillates between the figures of the virgin and the prostitute. Here, the extraordinary both emerges from the everyday and is more homogeneous than the latter. However, rather than being able to objectivize himself in the extraordinary world of the Ansaris, Sufi Baba can be seen as someone who becomes an object of ridicule and occasional awe for those who witness the drama.

THE CONCEPT OF WORK

While anthropologists and, in a more limited sense, sociologists have embedded work in a variety of structures and contexts[8] from which it takes its organization and meaning, conceptual constructions of the category of work parallel the type of society being studied. Broadly, issues of economic 'rationality' operative in a perfect market of standard products are the domain of the study of work in industrial societies, while the connections with kinship, religion, politics and so on are the exclusive preserve of those studying archaic or preindustrial groups.[9] A sociology of work has perhaps proved most fruitful in the study of the labour process, as evidenced in the debates Braverman's (1974) scholarship has

---

[8] See Wallman (1979) and Burawoy (1985).

[9] There are, of course, exceptions to this broad division. Geertz's account (1979) on the 'efficiency' of economic practices of the bazaar refer to the rationality of investing time to establish relationships of trust so that risks are minimized even when the environment is not conducive to the manufacture of standardized products. Further, Gudeman's (1986) recent study recognizes that models of livelihood are culturally produced and have a historical referent. Historians writing on work have begun to question models of economic rationality that anchor industrial society. A summary of these views is found in the introduction by Joyce (1987).

engendered, while the analysis of work in preindustrial societies has not consolidated on the study of Richards (1948). Whatever its merits, individual scholarship, in situating the category of work in industrial societies, with the emphasis on the labour process, and in non-industrial ones, with a corresponding stress on religious or kinship factors, ignores the importance of economic rationality and the scale of technology (in the case of non-industrial societies), and the influence of social structure (in the case of industrial societies).

When these two points of view are collapsed, the category of work is broadened to include virtually all social activity. Wallman (1979: 20) defines work as 'the production, management and conversion of the resources necessary to livelihood.' She identifies six such resources: land, labour, capital, time, information and identity, all of which have an economic and non-economic (social or personal) value. Schwimmer (1979: 287–315) suggests, contrarily, that work as a universal concept should be abandoned in favour of a comparative semiotic analysis. However, through the use of a particular method he advocates the reflexive projection of industrial societies onto non-industrial ones.[10]

The problem with Schwimmer's approach is its assumption that the concept of work can be decomposed into binary oppositions. This approach is problematic (as we will see in detail in chapters three and four) because it glosses over the polysemic nature of the symbolism. Schwimmer begins from a supposed core of what an economic model is supposed to be. He then derives other characteristics from this model. In one case the core is a rational choice in terms of the use of given factors, in another

[10] Focusing on a specific society of Papua New Guinea, he describes the concept of work with reference to two oppositions: exchange value as opposed to use value in terms of the amount of labour available and its utilization and whether the relationship between man and his work is 'alienated', or is one of 'identification' (1979: 295). Relating the two sets of oppositions, he distinguishes four systems of work concepts: (i) identification where use value is dominant; (ii) alienation where exchange value is dominant; (iii) alienation where use value is dominant; (iv) identification where exchange value is dominant. While the first two possibilities are congruent in that the derived concepts of work fit ideologies that reproduce the respective modes of production, the last two possibilities are incongruent because they are deviant ideologies (1979: 297).

it is a deviation from this choice. Consequently, this model employs a western category as the axiomatic core for understanding the data offered by other economies.

An alternative way of looking at work, adopted in this study, is to examine how the concerned community constitutes itself in a series of representations available in both practical action and conceptual models. The attempt, in other words, is to examine the category of work that is endogenous to the community. This means we have to proceed beyond the 'social context' of work — its embeddedness in various structures — and see how that context makes operational and is operationalized by work. In the process, the contours of the community become available. Further the dominant themes and issues inherent in work resonate in the ritual of circumcision and the biography of Sufi Baba.

When I say there is a resonance between various aspects of Ansari social life, I do not argue for a homology between work and non-work, though there is a continuity between the two. For example, through weaving and quilt making, the worker's body is constituted and classified. A similar operation is evident in circumcision rituals where the body of the novice is, in almost a literal sense, fabricated and socially legitimated. Similarly to heal the woman in pain, Sufi Baba first constitutes her impaired body, just as much as he reformulates his own in each such encounter. Such resonances are indicative of more general themes that link weaving and quilt making to the circumcision ritual and the biography of Sufi Baba. In all these cases we find a dialogical structure ordered in various ways. In the first instance, this ordering refers to the combination of discourse and practice. Within each discursive/practical domain we find the 'other' occupying a central place. This other is not merely a significant member of the domestic group, but can also be conceived of in a more abstract way. As the addressee of each discourse/practice, the other is a structure of possible genealogical and affective positions that an individual may occupy in his/her life. Simultaneously, in each of the domains the other exceeds the boundaries of the domestic group and shows how the Ansari community makes claims to an Islamic heritage. In this movement outside the domestic domain, the work of weaving perhaps supplies the most coherent cosmology.

Social Structure

Chapter two maps out the social structure of the Ansaris with reference to the place of work and the ritual of circumcision. The focal point is the household, which is part of the kinship and ritual organization of the community as well as the basic unit of production. The household embodies the relationship between determinate categories of people and these categories and material objects. These two types of relationships code the space of the household. Following local usage, the household is referred to as *chulah* (hearth). Here the basic kinship institutions of the Ansaris impinge on the division of labour of weaving and inform the ritual of circumcision. These institutions regulate the supply of labour to the weaving household as well as constitute the space of work. That is to say, certain spaces, for instance the *zanana*, are the exclusive preserve of certain categories of women, while the *mardana*, where the loom is located, is the domain of certain categories of men. These spaces are invested with social meaning through a series of movements enacted within them. The most important movements for our purpose concern the productive and reproductive life of the household. Through the act of weaving and quilt making on the one side and the ritual of circumcision on the other, the space of the household acquires a substantial identity.

In this sense social space is produced as the effect of work and of ritual. In turn both work and ritual are organized on the basis of the genealogical position an individual occupies in the household, the agnatic line and the community at large. Kinship positions within households are assigned a place in distinction to space. For instance, while the place of the male head of the household is exemplified by the act of making weft members of cloth in the work shed, the place of the female head is best crystallized in the quilt room and the courtyard of the zanana. In the former she stitches quilts; in the latter she is a significant actor in the circumcision of her child. In effect, place delimits the space of the household. The cardinal spaces of the household — the mardana and the zanana — have determinate places associated with them. The zanana is identified with the quilt room where three types of activity occur: quilt stitching, reeling the first bobbin for the weaving process and the consummation of marriage on the nuptial night. The circumcision ritual precedes the consummation of marriage in that a resemblance is established between the removal of the prepuce and the rupture of the hymen. Significantly, both these rituals are authorized in the zanana. Two types of activity are associated with the work shed: preparing the warp and the weft.

As mentioned earlier, the kinship institutions of the Ansaris organize the division of labour of weaving and provide membership for the circumcision ritual. These institutions reflect on the affinal structure of the community, the way kinship terminology is used in a system characterized by cousin marriage, and the conditions under which a household partitions from its parent dwelling. These three domains inhere both in the circumcision ritual and the labour of weaving by regulating access into the household. Consequently both work and ritual are influenced by the kinship structure of the community. In weaving this is most evident when labour is either solicited or contracted. In the circumcision ritual the kinship structure is most visible when we consider the genealogical place of actors.

Seen in the way outlined above, work and ritual may be apprehended through a 'social context'. This context determines the regulation and type of labour supplied to a household, the imperatives under which a household provides its labour to another, the way marriages between cousins (especially father's brother's daughter's marriage) alter relationships of work, and how because of the possibility of such marriages, the fissioning of the household becomes a necessity, the mode by which the newly-wed bride renegotiates her status as a member of her conjugal hearth and so on. For the ritual, the kinship context is seen in the genealogical position of actors present at the ceremony, the opening of the domestic group to determinate categories of people, the importance of descent and affinity in marking the novice's career in his community and the place of naming.

Thus the kinship structure of the Ansaris 'frames' work and ritual. I use the term frame as appropriate background. It establishes the setting for the execution of work and ritual but is not in itself representational. Against this background work has a number of meanings associated with it. Within the household it is considered a right fixed in accordance with the kinship positions that individual members of the household occupy. Outside the household it is thought of both as an obligation and a means of livelihood. For this reason, work secures a livelihood by accentuating a normatively oriented way of doing. Through work a weaver produces a commodity, but also participates in an exchange relationship given in the idea of gift. The gift relationship reaches its fruition in women's work.

Similarly, within the frame of kinship the ritual of circumcision has a number of meanings associated with it. From the point of view of the mother it connotes differently than from the point of view of the ritual specialist (the female barber). This, in turn, is different from the accounts of male witnesses of the ritual. Inside the domestic group the ritual is a necessary precursor to marriage, and outside it, the discourse on circumcision becomes a metanarrative by which the limits of the community are established. In the latter sense this discourse exceeds the domain of kinship.

If weaving is framed by the kinship structure of the Ansaris it is important to recognize that weaving acts as a framing device. The arrangement of the loom in the work shed provides cues which establish the significance of kin relationships within the domestic domain and the agnatic line. Also, in the act of making cloth the body of the worker is given expression. This latter theme runs through chapters three, four and five. The point is that the practice of making cloth both demarcates its area (the space of the hearth and the body) and invests with significance the work of making cloth for the shroud (*kafan*) for a member of the household whose death is imminent.

This type of practice is significant in as much as it, in Bourdieu's terms, owes its practical coherence to the fact that it is the product of a single system of conceptual schemes imminent in it. These schemes organize the perception of objects. They show how yarn and cloth are classified, place as direction is oriented (always in reference to the divisions of the body: up/down, left/right, front/back, differentiated/undifferentiated), time is coded (day/night, light/dark) and the agents of this action are invested with significance (found primarily in the use of kinship terminology). These schemes organize the production of practices seen through the gestures of the body. Such gestures are apprehended through the body techniques of weaving.

The circumcision ritual, too, acts as a frame that initiates the career of the male in his domestic group. The ritual makes available to the male novice significant members living outside his domestic group, as well as divides his body into distinguishable units. In this sense the ritual punctuates the time of the body and invests such time with social legitimacy.

The Body at Work: Technique, Finalization and the Labouring Body

The way weaving constitutes its world is developed in chapter three. In explaining the conventions that inform the production of cloth I will draw out the tradition of weaving. This tradition is understood, first, in the way time is coded and given content, and, second, in the mode by which the worker's body and that of the corpse is constituted. The coding of time is available from the use of three numbers — four, seven and twelve — employed in the act of weaving and invested with qualitative significance. The constitution of the body is given in weaving colours and designs and the relation of the body to the work process.

While this tradition is best expressed during the ritual occasion of making cloth for the shroud, it is also discernible, though in a diffused way, in the everyday world of weaving. The everyday and the extraordinary are linked through the techniques of work. Using Mauss (1973) as my point of departure I argue that techniques of weaving establish relationships of equivalence: between the body of the worker and the instruments of work, and between word and object. In effect, such techniques are instrumental in that they refer to the craft of fabricating cloth. This craft is based on the learned repetition of body gestures. Techniques are also expressive since body gestures define certain categories of workers. In this way, techniques of weaving are a mode by which the community frames its individual members. In the above sense, technique in the usage of Castoriadis (1981), 'stands for' and 'serves for' something.

What is important in the discussion on technique is the relation between weaving and the body. To the extent that techniques codify attitudes, postures and gestures of the body at work, they fabricate social relationships. In this sense the worker's body is delimited and articulated in its actions. Every technique, as far as weaving is concerned, is inscribed on the body. From the work of children to making cloth for the shroud, weaving techniques appropriate the body at work. Individual workers are transformed into signifiers of rule, as the tradition of weaving is incarnated 'as flesh'. In this way, too, individual workers become members of the community.

At the same time, a second operation is discernible, one that runs parallel to the first. This operation shows how, through the fabrication of cloth for the shroud, the corpse (body) is invested with an ideal corporeal and non-material integrity. The attempt at reinvigorating the corpse by infusing it with carnality and spirituality is achieved by using the four elements of nature — earth, water, fire and air — through the agency of four categories of workers. Each of these workers imprints the cloth for the shroud with a distinctive signature. I use signature in two ways: it points to a distinctive style, exclusive to a certain type of worker engaged in making cloth for the shroud; second, this authorship is carried out by reference to a code which establishes normative limits. For instance, the wife's brother or the mother's brother (this worker is known as *mama*) of the male head of the household cures yarn as his service of making cloth for the shroud. While engaged in this work the mama represents the features of earth (a dormant femininity) and water (humidity). After the yarn is washed he typifies a combination of fire and air, manifested as dry heat and an awakened femininity.

The style of each worker is made possible by means of instruments which codify the body gestures of the worker. In this way, each style produces practitioners. The relation between the worker and his instruments creates a field that authorizes social action. It is because the mama bathes the yarn in rice water, then dries it by holding it up on two bamboo poles and places lighted lanterns close to it, that he is characterized by a signature. This signature is not available to the bride who cures yarn in everyday life because as a worker she can be substituted. Her conjugal household may financially contract the labour, a contract that is denied when cloth for the shroud is woven.

There are, however, determinate connections between the everyday and extraordinary acts of weaving. These links are not limited to the use of the same techniques but also refer to the type of personnel engaged in the various stages of weaving and the coding of space. The mother's brother (hereafter MB) and the bride (*dulhan*) carry out tasks associated with sizing. Both stand in an affinal relationship with the concerned household. Their work space is fixed in the mardana. The bride, unlike the MB, renegotiates her status in her movement to the inner part of her conjugal home. The MB dries the yarn in the zanana.[11]

The clearest link between the everyday and the extraordinary is found in the coding of time. In everyday life the act of work divides the day into four parts. Each part has a category of worker and a segment of the ritual calendar associated with it. When cloth for the shroud is made these divisions are formalized. In addition, an explicit significance is attributed to the seven days of the week. The esoteric conception of time, faintly discernible in everyday life, achieves fruition in the extraordinary world when the corpse becomes an ancestor. In analysing the loom as 'pole' (*qutb*) I argue that this time is transhistorical. This chapter suggests that the organization of the space of weaving, the imprinting of the four signatures on the body of the corpse and the oscillation of weaving from night to day, are formalized in the relation between the light and the dark.

By focussing on the relationship between the light and the dark both the tradition and act of weaving are finalized. By designating the act of weaving and organizing its construction finalization tells a 'story'. It highlights the combination of formal operations: the significance of work done in certain areas of the household, the importance of categories of workers, the necessity of the temporal structuring of weaving and of learning body gestures while at work.

---

[11] The dry yarn is reeled into bobbins by the woman head and the other women of the household, or, when cloth for the shroud is being made, by all the women of the extended family in the zanana of the household. The work is not only collective but also points to the role of the woman as creator and nourisher. Warping is the domain of the younger brother or son of the male head of the household, and of the FBS when cloth for the shroud is fashioned. The domain of work is in the work shed. A display of learned skills is necessary for both workers and rests on an equilibrium of colour and design. Both of them are apprentices. Finally, weft members of cloth are made by the male head of the household, or of the agnatic line, in the work shed. Here, work does not only result in the production of cloth, but also in the reproduction of the community.

Finalization points to the coherence, instrumental and expressive, of the craft. Coherence refers to the conventions informing the production of cloth and is studied in two ways: in the sequential events leading to cloth production and their structuring. In everyday life these referents are the personnel involved in cloth production, their relationship to the instruments of work, the location of work and its temporal rhythms. These events are structured in accordance with the amount and type of cloth that is to be manufactured. For this reason it becomes important to see the type of colours and designs used. In making cloth for the shroud the sequence of events and their structuring follow the pattern of the everyday. The difference is that the corpse is centrally located in the work cycle. Because of this weaving acquires an esoteric character and explicitly reproduces the tradition of the community.

PRACTICAL TAXONOMIES AND DIALOGUE

Finalization shows that the tradition of weaving, as it unfolds in the everyday and extraordinary life of the Ansaris, is a coherent act. The concept of dialogue argues that this coherence is not that of a closed structural unity, which only promotes understanding from an internal perspective. Chapter four examines weaving from the aspect of material culture by focussing on two points: the constitution of the loom as a material object and the relationship between this object and the community of weavers. Here I argue that craft objects are not amenable to a semiotic operation determined by the Saussurian method of studying language. Instead of privileging an ideal and logical model of language which emphasizes binary oppositions (langue/parole, signifier/signified, syntagmatic/paradigmatic, denotation/connotation), I suggest that the loom be read in terms of the practices to which it is put, for the meaning of work changes depending on the practice. There are four such practices where the loom is the focal point: (i) everyday production of cloth; (ii) production for the shroud; (iii) initiation ceremony of male children into weaving; (iv) transmission of the loom.[12]

[12] Chapter three discusses the everyday and extraordinary world of weaving, while chapter four localizes this division in the order of the loom.

Using Bourdieu's notion of practical taxonomies I argue that the four practices operate simultaneously in two directions. The first two refer to the naturalization of cultural practices by which the boundaries of work on the loom are established and reproduced. Such boundaries point to types of workers who can be weavers and those who can never weave. In effect, these boundaries order the division of labour of weaving. Everyday production and production for the shroud authorize the division of labour of weaving. This authority establishes a temporal continuity by highlighting a memory found in supplicatory prayers (*du'a*) the concerned weaver recites while working the loom. The function of these prayers is that they integrate the tradition of weaving with the present moment of weaving.

The latter two practices, the initiation ceremony of male children into weaving and the transmission of the loom, refer to the processes of socialization by which the boundaries, established by the first two practices, exceed the individual life of the weaver and provide the necessary conditions for the perpetuation of the community. The recitation of prayers in the latter two practices establish contact with the work tradition through the implanting of memory in a place. In the initiation ceremony this implanting is in the domestic domain, while in the transmission ceremony it is localized in the work shed. Through this recall progeny emerges as a value, and it is recognized that the status of the head weaver is transitory. To this extent, the work tradition, as memory, is fashioned by external circumstances. For this reason, the tradition of weaving cannot be thought of as unitary.

Common to the four practices is the recitation of supplicatory prayers, found in the *Mufidul Mu'minin*, a text used by the weavers of the area in varying ways. This text is not merely a primer on weaving, but is, in Izutsu's terms (1964: 18–24), a *kitab* in the Quranic sense. I argue that the text is framed by the term *nurbaf* (weaving of light). It differs from a strictly Quranic frame in that it does not establish a revelatory relationship between God and man, but a supplicatory one between man and God. This is suggested because the text, in disaggregating the

---

Here, a certain ethnographic repetition is unavoidable, but the point of the argument is different. It is not so much to show the tradition of weaving as to argue that this tradition is unfinished and open-ended.

loom into its component pieces associates each piece with the recitation of a prayer and then decomposes each piece into its phonemes in a way that sounds embody virtue. In this way the loom is framed in language.[13]

Given that the loom is constituted in language, it is possible to study this language in terms of the Saussurian semiotic system. However, I argue that the loom has fluid boundaries. While in everyday weaving three or four sets of prayers are recited, in the transmission ceremony almost the entire text is read out and certain portions of it commented upon. More significantly, the prayers establish a communicative relationship both between man and God, and man and man. These prayers, thus, are utterances in Bakhtin's use of the term and not sentences or propositions.

In effect, linguistic matter constitutes part of the utterance. It has a non-verbal part corresponding to the context in which the utterance is made. This context comprises a temporal and spatial horizon common to the addresser and addressee, the reference of the utterance and its evaluation. Because these prayers as utterances establish a relation between the addresser and addressee they constitute the other. In this constitution of the other the time of weaving is instituted. Three notions of temporality are expressed through utterances. As mnemonics they establish a continuity with the work tradition so that time is continuous but relocated. Second, utterances mark out thresholds where time is instantaneous. Third, because they are recited and repeated utterances point to the singularity of time.

The other is constituted by prayers serving as dialogue. As mentioned earlier, these prayers, directed at determinate members of either the domestic domain or the agnatic line, fix the position of the addresser and addressee. In everyday weaving the other is a potential successor who, with the passage of time, takes the place of the utterer. In weaving for the shroud the other can never be a competitor because it is for his shroud that cloth is being woven. In the initiation ceremony the other is an initiate into the tradition of weaving, while in the transmission ceremony the other inherits this tradition. As far as the recitation

---

[13] In the discussion of the cosmology of weaving a similar typology operates for individual workers and the corpse.

26 • *Work, Ritual, Biography*

of utterances is concerned, the first three practices emphasize reported speech, and the last authorial speech.

The dialogue, in the sense of a communicative relationship between addresser and addressee, is one where transmission and reception occurs simultaneously, not sequentially. Further, this dialogue is part of an action which combines verbal and non-verbal gestures. In this sense, to work is as much to verbalise as it is to express through the movements of the body.[14] Each practice uses the same utterances differently. Consequently the determination of the meaning of work is dependent on the meaning of the practice. Thus, in everyday weaving work is constituted as a right; in weaving for the shroud work is obligatory; in the initiation ceremony work is construed as substitution because the addresser substitutes for the addressee; in the transmission ceremony work is constituted as gift.

POLYPHONIC DIALOGUE

The above dialogue is directed both towards a referential object (the operation of the loom) and a received discourse (in that it occurs in the *Mufidul Mu'minin*). Here, the addresser as stylizer uses another's discourse precisely as other. While reciting prayers mentioned in the text, the worker does not embody the original speaker, but is using his speech in a normative way. Thus, the addresser makes this discourse double-voiced. The recitation of the addresser establishes a relationship with the written word as well as with a formalized technique of doing. In quilt making, on the contrary, the relation between addresser and addressee is not mediated either by the written word or a formal technique of doing. In fact, the quilt maker as author of a particular discourse about women works with materials that are heterogeneous and external to her work. Chapter five argues that quilt making establishes a relationship of resemblance between the quilt and the quilt maker and between different workers.

It may be argued that quilt making, undertaken by women of the household, and during marriages, by women of the wider

[14] Implicit in the classification of the loom is a division of the body into up/down, left/right, front/back. This division recurs repeatedly in the process of weaving and quilt making.

community, is determined by the tradition of weaving since the latter is based on the presence of a cosmology and the written word. It is an authoritative discourse which fixes the place of women workers (as a type who can never be weavers). This chapter makes the opposite point. The quilt can be neither produced nor read in the way woven cloth is because the nature of materials composing the quilt is different from that of weaving. Codified techniques of quilt stitching do not exist. This work is not premised on a structure of conventions which establish an impermeable division of labour. The dialogue inherent in quilt making is part of ordinary language while in weaving it is formalized as du'a. As finished product the quilt is used by the household, or is a gift made to a daughter who is getting married. Finally, in the act of manufacture, the producer asserts her status as the female head of the household and not as a worker involved in weaving chores.

Quilt making is divided into two domains: the everyday and extraordinary. The quilt used in everyday life, composed of discarded rags, is a patchwork quilt. The extraordinary quilt, stitched when a woman is to be married, is embroidered and fashioned out of two or three pieces of fine quality cloth. The patchwork quilt is stitched by a single woman of the household and the embroidered one by women of the extended family. The techniques of making these two types of quilts are different.

When I say the relationship between the quilt maker and the quilt is one of resemblance I do not argue for a mimetic discourse and practice. Furthermore, a direct authorial discourse and practice of quilt stitching is not premised on the presence of authoritative points of view and stylized ways of doing. Resemblance points in two directions. It stylizes another worker's way of doing in the direction of the style's own specific tasks by making these tasks conventional. Following from this, the act of resemblance, in refracting the original worker's intention, introduces a semantic change via the resources of improvization.

The above notion of resemblance is seen in the use of the embroidered quilt. Before women begin their embroidery the type of designs to be embroidered is decided. Since a formally authorized code according to which designs are embroidered does not exist, distinctive motifs that were made in the past are evoked to justify those in the present. The reference point of this evocation is known as the *lal kitab* (red book), an ad-hoc compendium of various events that happened in the past. The red book is not a material object but a collective storehouse of memories usually used to legitimate a particular point of view. It is both strategy and resource, much like Bourdieu's habitus. As far as the embroidered quilt is concerned, the use of designs mentioned in the book become a reference point as well as an authorized way of embroidering designs. Once the embroidery begins each worker demonstrates her skill by varying the design. In this way resemblance constitutes the present as the place and time of licence, but always elusively.

Resemblance is expressed differently when the patchwork quilt is stitched. Here, through her work the worker is engaged in self-portraiture. This is understood through a semiotic analysis of the quilt. To this end I discuss the type of rags used in making the quilt, the relation of the body to the act of stitching, the type of colours and designs employed in making it and the nature of the dialogue the worker engages in. This dialogue, I argue, is polyphonic. One dominant theme emerging from it is the expression of pain. This pain reflects both the arduousness of the task and the condition of the worker.

The polyphony of dialogue points also to a meaning of work which resists integration in a unitary sense. Unlike weaving, which proceeds according to a systematic way of doing and where the meaning of work emerges from the type of practice it is inscribed in, in quilt making the meaning of work is inscribed in the experience of each worker, giving rise to varied nuances. Given its heterogeneous quality, it may be argued that the meaning of work derived from quilt making is incapable of being generalized.

Quilt making, however, is informed, just as weaving, by discernible general features. First, both the patchwork and embroidered quilts suspend the hierarchy instituted by weaving by reinforcing and reiterating the privileged role of women as quilt makers. In the embroidered quilt the distance between workers is suspended, leading to free and familiar contact among them. Second, quilt making emphasizes the division between men and women. This division is found in reserving a domain of work exclusively for women and as far as the patchwork quilt is concerned, in its very organisation. Third, the collectivity instituted by work for the marriage quilt is bound in its polyphony of voices. Fourth, work on the embroidered quilt is characterized by improvisation. Finally, on marriage, quilt manufacture establishes an elaborate gift relationship.

RITUAL AND BIOGRAPHY

While the chapters on work draw out the structures and institutions within which weaving and quilt making are situated and which, in turn, authorize such structures, the subsequent two chapters attempt a reading of the circumcision ritual and its everyday discourse on the one side and the relationship of a particular individual to the Ansari community, on the other. Focusing on specific cases these chapters show how the domestic and extended kin-group inhere in the lives of individual members (chapter six), and the resistance to the norms of domesticity and worldly kin relationships in the life of Sufi Baba, whose position in relation to the Ansari social structure is ambiguous (chapter seven). In both these chapters, the body of individual actors becomes the site on which the social structure is both impressed and resisted. If the main effect, and, one might say, affect of the circumcision ritual is to prepare the male body for entry into the domestic group, in the case of Sufi Baba the presence of the male Ansari community is sought to be resisted to the extent that through bodily expression he does not produce those orthodox signs by which he can be classified according to the divisions of the Ansari social structure.

THE GESTURAL AND GRAPHIC BODY

In shifting the focus of study outside the domain of work, the sixth chapter analyses the significance of circumcision for the Ansaris. The chapter is divided into two parts. The first discusses the ritual of circumcision, called *khatna* (to cut) by the Ansaris, while the second draws out the contours of the fact of being circumcised, called *musalmani*. In showing the semantic range of khatna and musalmani, the chapter shows the relationship between the everyday (the fact of being circumcised) and the extraordinary (the ritual).

Briefly, the Ansaris circumcise their male offspring between the ages of two and six. The operation is carried out in the courtyard of the zanana of the novice's father's household. After the operation the prepuce is buried under the nuptial bed, located in the quilt room. Space, here, is oriented in two ways. The courtyard of the zanana resembles the wall of the mosque facing Mecca (in this case westwards). The entire courtyard, cleared of all impedimenta, is characterized by a sparse look. The quilt room, in contrast, is decorated with flowers. The nuptial bed is prepared and a little perfume sprayed on it. An earthen pot placed on an elevation near the bed, is broken after the operation.

The zanana, in Bourdieu's sense, acquires a direction and orientation. In the courtyard all action is directed towards the western wall, while in the quilt room the basic contrast is between the pit where the prepuce is buried and the elevated earthen pot. Just as for the ritual of cloth manufacture for the shroud and the marriage of a daughter of the household, during the circumcision ceremony the home opens its boundaries to members of the extended household (*kumba*) and the circle of classificatory relatives (*bihaderi*). These members include the boy's father's male agnates and the boy's mother's brother. The female barber as ritual specialist plays an important role in the ceremony. The mother prepares her child for the operation, while the barber removes the prepuce. After the operation the eldest male agnate present in the ceremony whispers the boy's formal name and the *azan* in the novice's right ear. Here women execute acts that are primarily gestural while males carry out acts that are primarily verbal. Together these two acts, I will argue, compose the body of the novice.

The circumcision ritual shows how the novice's body is classified and made substantial through the inscription of the ritual wound. At issue is the nature of this ritual. The sixth chapter argues that the ritual is informed by the idea of 'biunity'. I use the term biunity in three ways. By combining the characteristics of male and female, the ritual, far from separating the sexes, premises a unity between them. The term also signifies a relationship between the physical and social body, where the social legislates and substitutes for the physical and a relationship by which the circumcised enter both the social life of the domestic group and the community of Islam. Each of these aspects of biunity is made operational in either a gestural or graphic mode or in a combination of the two.

The emphasis in the ritual is on the body as a referential object. There is another aspect to circumcision, one where the body is absented from thematic awareness. Faintly discernible in the ritual, this theme becomes dominant in everyday discourse. Unlike khatna, which refers to the body of the novice, in musalmani the referent becomes an ornamental inscription in the pronouncements of men. This is seen most clearly in the selective appropriation of the meaning of khatna in everyday discourse, one which highlights the verbal aspects of the ritual. Yet, a certain meaning of khatna always resists integration. This meaning refers to the classification of the body and its career in the domestic group. Furthermore, the two terms, khatna and musalmani, are not hierarchically ordered. They are linked through the articulation of a collective memory. That is to say, both the discourse of musalmani and the ritual of khatna employ procedures by which the body is valued in incorporeal terms. In the case of the ritual the valuation is found in the enunciation of the azan and the boy's formal name, while in everyday discourse we find the emergence of a master narrative, one where the parameters of the Muslim community, as conceived by the Ansaris, are established and simultaneously the limits of being Muslim also stated.

Bourdieu (1977: 225 n. 56) says that khatna is a purificatory cut that protects the male from the dangers of sexual union. Furthermore, circumcision is a type of practice where the young male is symbolically torn from the female world. Here, women are expressly excluded. For these reasons circumcision is a second birth, a male event enacted outside the house, in a space primarily associated with men — the agricultural field. Bourdieu finds a homology between the act of circumcision and that of ploughing. The uprooted field, cleared of all traces of life and vegetation, is equivalent to the incorporation of the circumcised boy into the community of men (1977: 135). Given this correspondence, Bourdieu argues that the circumcision ceremony, the ploughing of the field and a range of other practices are generated by the habitus. This habitus divides space, both of the body and external geography, into its constituent units, gives time substance and thus provides coherence to the various practices of the group. Circumcision, as I analyse it, can be arranged within Bourdieu's scheme of practical taxonomies. This is found in the relation between the circumcision ceremony and marriage, on the one hand, and khatna and cloth manufacture for the shroud, on the other.

The view that circumcision is part of a structure of practices, organized in a more or less coherent manner, is one axis for anchoring the Ansari community. It is important to recognize, however, that musalmani is discursively formulated in a way that all those who speak in its name make claims to an Islamic heritage. This discursiveness shares a strong similarity with the ritual of khatna. In the ritual the boy's name and the azan are dialogically related to the inscription of the ritual wound on the body: with the wound the boy enters into the productive and reproductive life of his domestic group; with the azan the liturgy is inscribed on his body and individuated through the enunciation of his formal name. In this sense the dialogue between the verbal and the gestural recognizes the simultaneous difference between the two. Implicit in this dialogue is the disciplining of the novice's body, evident in the importance placed on various hygienic practices.

The discipline and surveillance of bodies is strongly marked in the discourse on musalmani, but in a way that is different from khatna. Here, the techniques of weaving are imbued with a strong morality. Such techniques refer not so much to the skill of the weaver as to the view that he is a 'good Muslim'. The third chapter shows that a skilful weaver is known as someone who is 'strong of speech'. Musalmani encompasses this statement by arguing that a wise weaver and good Muslim practices weaving under conditions of privation precisely because by being a good Muslim he is witness to the sacrifice and suffering of the archetypes of Islam as conceived by the Ansaris: the Prophet, Husain, Sis Ali and Ayub Ansari. The latter two are identified by weavers as their genealogical fathers. The first act of pain a Muslim feels is when he is circumcised. In this sense pain is thought to hark back to the suffering of the archetypes. For this reason, pain is built into the definition of the community and is seen to be synonymous with belief. Musalmani, belief and pain form a set.

The association of musalmani with pain and suffering allows for the representation of the Ansari community as one formed in hardship. The fundamental difference established here is between

Muslims and Hindus. The latter can never be good weavers and by implication good Muslims because they are inadequately experienced in pain. The discourse of musalmani, then, is constitutive in that it establishes the definition of the Ansari Muslim community. But this discourse is also 'reported speech' in Bakhtin's sense because it is laid out in its telling. This telling is the way the authority of men is articulated in relation to the gestures of women.

INNER VOICE AND OUTER SPEECH

The sixth chapter shows that the speech of men provides legitimacy to the gestures of women by incorporating their practices within the norms established by musalmani. The seventh chapter shows how such authority is interrogated, at least implicitly, by Sufi Baba. This interrogation is available in the way he constitutes his body in relation to women in distress and during the most important festival of the Ansaris, known as *Chahullam*. In the first case, Sufi Baba interacts with women in pain with the object of curing them of their affliction, while in the second, as a member of the Ansari community, he enacts in a dramatic fashion the roles of clown and spokesman. He also presents a third persona in this festival, where dressed as a woman, he attempts to solicit customers for prostitutes. I will explore the relation between these two types of practices. Accordingly, the chapter is divided into two parts. The first discusses Sufi Baba's therapeutics and the second locates him in the festival of Chahullam. His relationship with women in pain is part of his everyday life, while his participation in Chahullam occurs once in the lunar year.

I argue that these two types of practices can be understood as a 'double voiced' discourse, in that Sufi Baba explicitly distinguishes between his private relations with Ansari women in pain and his public persona in Chahullam. He attempts to resolve the distinction between the two by a contemplative mode of worship which focuses on the figures of the virgin (for women in pain) and the prostitute (during Chahullam). As figures of worship, the virgin and the prostitute are formulated in a way that allows him to strategically interact with Ansari women and men.

In his relationship with suffering women Sufi Baba authors a cure and is responsible for reading signs of impairment in his own way. As far as impairment is concerned, a cure, according to him, can only be found if one is pure in one's heart. The untouched virgin comes closest to this notion of purity. Contrarily, in Chahullam, Sufi Baba is inscribed with the signs of the Ansari community and is formulated by the collective in a way that agency is often denied to him. Here, he is the subject of signs. Thus, we see that his therapeutics is premised on a condition where, through his inner voice, he authors signs, while in his role as clown and spokesman he is authored by the signs of the community. In the latter case he is often subject to verbal ridicule, but it is here that his behaviour and sharp repartee can be seen to critique the orthodox definitions of the community. This critique becomes all the more effective because Sufi Baba, though an Ansari and recognized to be thus, does not belong to any kinship institutions save the *biraderi*: his house is not situated within a kumba and he does not claim or owe allegiance to any bihaderi. As far as the biraderi is concerned, Sufi Baba claims to be a descendent of Ayub Ansari. More significantly, in venerating the figure of the prostitute he is able to rationalize the ridicule of his community. Yet, even here he retains an autonomous domain of practice by saying that he solicits customers because he himself is God's prostitute.

The autonomy he arrogates to himself is fashioned in two ways. In his relationship with suffering women he calls himself a healer of pain. The initial reading of cure is premised on the classification of a balanced body into the four elements of earth, water, fire and air. The impaired body is one where these elements are not in equilibrium with each other. Sufi Baba's cure is premised on his ability to read this classification, but in a way that he becomes a partner with his patient. The ability to take on the pain of his patient is the basis of his therapeutics, one where a cure is found depending on how pain manifests itself in his own body. In this sense his body produces a cure. Because Sufi Baba authors signs and because no other can replicate his experiences he constitutes his body intransitively.

Sufi Baba's body is constituted in a second way during Chahullam. In this festival he is forced into a condition of self-alienation from his avowed vocation as a person of contemplation. This is clearly marked in a dance he performs during one of the nights of the festival. Dressed in a combination of male and female clothes, he is questioned about his role as healer. The questions reflect on the feminine aspects of his performance, but in a mocking way. Sufi Baba's repartee, equally reflective on the body, attempt to create a zone which recognizes the female body. However, in his alienation from the community, and eventually from himself, he attributes to himself a meaning which is formulated in reference to exile. He expresses this exile as the loss of both an external prophecy and an explicitly gendered identity. The absence of a gendered identity is questioned most acutely when he mediates between customer and prostitute, particularly because by symbolizing the words and gestures of prostitutes he is thought to be of ambiguous sexuality. He justifies his mediation by arguing that in embodying the image of the prostitute, he expresses a desire that can be satisfied only with the recognition that the beloved manifests divine will. In any case, the exile that he speaks of is one where his body is invested with significance by the community. For this reason, his body is open to multiple meanings and is constituted transitively.

In effect, Sufi Baba's body is composed through two inscriptional practices. In the first, his body produces signs, and in the second, his body is constituted under the authority of the community. He attempts to bridge the two by a mode of worship that embodies the figures of the virgin and the prostitute. This embodiment is an act of 'imagination' in the way Castoriadis (1987) uses the term. He argues that every society conceives of itself as a productive and creative imagination manifested as 'doing' and 'representing' (1987: 127). In this conception every society defines and develops an image of the natural world and makes of it a signifying whole which rests on the order of the imaginary. In the context of the life of Sufi Baba this imagination is constituted as an active and passive imagination given in the figures of the virgin and the prostitute. In embodying these two images he thinks of himself as feminine: with the virgin his body is a receptacle and with the prostitute it is an agent responsible for initiating a desire for the beloved.

Chapter Two

Household, Kin, Work and Ritual in Barabanki

Treating domestic space as the initiating point of my inquiry into the working and sacred life of the Ansaris, I delineate the location of various kinship institutions in the organization of work and ritual within the household. I will argue that the household functions as part of the organization of production and is simultaneously an important element in the kinship and ritual structure of the weaving community.

I will first describe the basic kinship institutions with particular reference to the domestic hearth of the weavers and the place of weaving within it. I will show that weaving divides both the space and time of the household and the labour of the domestic group in accordance with gendered and generational positions. In restating the genealogical place of individuals, weaving locates them within the work cycle of the household. This work cycle is incomplete if limited to the household. Its full significance is realized when cloth for the shroud (kafan) is made i.e. when the community of weavers living outside the household make cloth for a member of the household whose death is imminent.

The basic kinship institutions of the Ansaris play an important role in the circumcision ceremony of the young male child. Just as for weaving, in the case of the ritual the domestic group and the agnatic line mark their presence on the body of the novice. This mark plots the career of the novice in the domestic group and also makes available to him the community of colleagues of his generation living in households other than his own.

The kinship and marriage practices of the Ansaris in relation to work and ritual are explained with reference to each of the three sub-areas of terminology, marriage and fission in the domestic group. The basic institutions of the Ansaris are the hearth (chulah), an extended family usually living in contiguous hearths (kumba), a circle of classificatory relatives living outside the kumba (bihaderi), and the brotherhood (biraderi). Before I describe these institutions, I will present my household schedule.

HOUSEHOLD SCHEDULE OF THE MUSLIMS OF WAJIDPUR

Ansaris	Household	Other Muslim Groups	Household
Bharaichiya	21	Khan Saheb	42
		Shaikh	14
Purabiya	18	Dhuniya (Cotton Carder)	11
Shaikh	17	Nai (Barber)	18
		Kasai (Butcher)	14
Faizabadi	16		
		Fakir (Mendicant)	20
Multani	15	Kabariya (Green Grocer)	47
		Mirasi (Musician)	30
Madari	14		
Muhammadi	12		
Total	**113**		**196**

I take a household to mean a dwelling which has a hearth and a loom. The Ansaris refer to a complete household as chulah. By this is meant that a household is composed of three generations of inhabitants, a fully functional hearth and an ancestral loom. In this sense, a household comprises social relations and property. I use the terms household and hearth as synonyms.

KINSHIP INSTITUTIONS: THE HEARTH (CHULAH)

The average membership of an Ansari hearth is approximately 6.5 people. Membership is determined agnatically in the case of males, and affinally in the case of women. Disregarding procreation, new entrants into the hearth are almost always women. The exception is the husband living in his wife's father's house (ghar jamai), but this again depends on the kin relationship the ghar jamai shares with his wife's father's hearth before he marries her. Both men and women may leave the household. In the case of men exit occurs typically under two conditions. In the first, the man may migrate to a city, usually Bhiwandi in Maharashtra, in search of a relatively well-paid job. Most hearths have in the course of their life at least one male member working in Bhiwandi. A second type of exit occurs when three generations live in the same house. Here the hearth partitions in the second generation, with the migrating male establishing his own dwelling. This partition is local, confined not merely to the same village, but often the same kumba. Among women, exit occurs through marriage, in either the second or third generation.

The hearth is the largest exogamous unit. In a formal sense, rules of exogamy are provided by Muslim law. Ansaris say that a person may not marry: (i) a sibling or a step-sibling; (ii) a descendant of a sibling; (iii) an ascendant of a sibling; (iv) a wife's mother and, during the lifetime of the wife, her sister. Sibs are of two types: those descended from the same parents and those weaned on the same woman's milk. Exogamous groups are important because for each individual a different circle of kin are brought into focus in the prescribed exogamous circle. Exogamous rules thus define relationships between individuals, but do not separate them into kin groups. Kin groups are defined by the norms of endogamy, and the institutions that fix these norms are the kumba, bihaderi and biraderi.

Members of individual hearths form the core working group for weaving. Weaving itself is divided into four stages: yarn curing or sizing, reeling bobbins, warping and preparing weft members, or what I call 'wefting'. Ideally, work for each stage is supervised by different hearth members. While women supervise the first two stages, men control the latter two. Further, between the first and second stages, as between the third and fourth, the overseers are distinguished by the generations they occupy. Thus, while the son's wife coordinates sizing, the woman head of the hearth directs reeling. Similarly, while warping is the province of the son/s, wefting is both controlled and performed by the male head. Weavers say a fully formed hearth is composed of three generations, while one of two generations is in the process of formation. The latter is seldom referred to as chulah. It is called *ghar*. It takes a minimum of workers belonging to three generations for a house to satisfy its weaving requirements without contracting labour.

The mother of the novice is a significant actor in the circumcision ritual. Together with the female barber she works on the body of the novice, while the male circumcised members of the chulah, kumba and bihaderi form part of a group of witnesses. Work in the ritual, as in weaving, is ordered along gendered lines: the mother physically prepares the body of the novice, while the witnesses authenticate this work by providing verbal legitimacy to the mother's labour. In the case of a household composed of three generations the task of preparing the novice is often conducted under the aegis of the eldest female.

Just as weaving activities of the hearth are differentiated by gender, so also the spatial organization of the hearth is bifurcated into a domain of men (mardana) and of women (zanana). These domains acquire significance through the work performed in them. The mardana comprises an elevated courtyard covered by a straw roof reserved for children of both sexes and married and marriageable males. Male guests, if they are not consanguineal relatives and/or of the mother's brother's (mama) household, are entertained in this courtyard. The mardana also houses the work shed (*karkhana*), where the loom and the warp beam are kept. This is the most exclusive part of the mardana and any intrusion while work is in progress is resented.

The zanana is the space of the womenfolk of the house. It includes an uncovered courtyard, within the precincts of the dwelling, reserved exclusively for married and marriageable women. Male children, until they have reached the age of puberty, are allowed access to the zanana. A part of the zanana houses the quilt room (*du'lai kamra*), where the first bobbin, called *kunda*, is reeled. The nuptial bed (*khatiya*), too, is located here. On the marriage of a daughter of the hearth, women of the kumba congregate in the courtyard of the zanana to stitch quilts. In such instances, this part of the hearth is zealously protected by women against encroachment by married and marriageable males.

If the mardana is the place of weaving, the zanana is the place, par excellence, of the circumcision ritual. During the ritual the courtyard of the zanana and the quilt room are marked areas. The courtyard is made to resemble a mosque while the quilt room is decorated as the room of the nuptial night. The nuptial bed is prepared and a little flour and perfume sprinkled on it.

As far as the productive life of the hearth is concerned, its spatial division into male and female zones corresponds to the distinction between quilt making and weaving. Quilt making, undertaken by women of the hearth and during marriages, of the kumba, occurs in the zanana. Males provide their services for quilts for everyday use, but are absent when quilts are stitched for the bride's trousseau. As finished product the quilt moves from the zanana to the outside of the hearth, where it is washed and left to dry. The operation of the loom, the exclusive preserve of men, occurs in the work shed. Women participate in the weaving process for sizing and reeling. When cloth is made for the shroud women participate only in reeling. This work is performed in the zanana, where quilts are made, or in the courtyard of the mardana, where quilts are left to dry. Finally, while cloth for everyday use is marketed, quilts made for everyday life are meant solely for the use of hearth members. The focal point of the zanana is the quilt room, while that of the mardana is the work shed.

Work Shed (Karkhana)

Work in the work shed is divided into two types of acts, performed sequentially. First, bobbins are mounted on the warp beam and warp members of cloth are fashioned by drawing the threads from the beam to the loom. Second, weft members are formed through the operation of the loom. Warping and wefting are executed here. The work shed is the place of the male head of the hearth and, when cloth for the shroud is made, of the agnatic head living within the same kumba. Other males must pass through the three prior stages before they exercise any right over the loom of the hearth.

The work shed is closed to the outsider, who may be, depending on the occasion, an outsider to the hearth or the kumba. The outsider cannot participate in any of the two activities occurring within the work shed. When cloth for the shroud is made, the work shed becomes a hermetically sealed unit to the extent that not even the head of the hearth is allowed entry. Here, the work shed is thought to be a place of worship. A weaver told me that when cloth for the shroud is made no place is closer to heaven than the work shed, perhaps echoing the Kaaba's unique position. The work shed has a special relevance for the circumcision ritual. The green cloth which drapes mother and child after the removal of the child's prepuce is made by the hearth's male head, preferably on an ancestral loom. As in weaving cloth for the shroud the work shed is the exclusive preserve of the weaver, but this time of the hearth's head. Further, no agnate living outside the hearth participates in any of the other three stages. The work shed is not considered a place of worship possibly because that status is reserved for the courtyard of the zanana.

Quilt Room (Du'lai Kamra)

The quilt room lends itself to the purpose of everyday weaving if quilts are not being made at the same time. During weaving children often work here, especially if labour is contracted for reeling. Just as the work shed is the preserve of the male head, so also is the quilt room that of the woman head of the hearth. Within the confines of this room the woman reels the first bobbin for everyday weaving. During this period no hearth member, either male or female, is allowed to disturb her. When a woman of the hearth is getting married the quilt room offers a structurally reversed contrast to the work shed when cloth for the shroud is being made. The contrast is not merely between one man working on cloth, and many women working on the quilt, but also in the type of membership to these activities. In the case of the work shed only one type of male makes such cloth, whereas, for the quilt room, membership is open to all Ansari women of the kumba except widows. Men are excluded from the quilt room and any intrusion is greeted with ridicule and verbal abuse.

Part of the circumcision ritual is expressly connected with the quilt room in particular, and marriage in general. After the operation the excised part is buried under the nuptial bed and an earthen pot is broken in the room. Given the details of the ritual we find a discernible connection between the removal of the prepuce and the deflowering of the virgin on her nuptial night. Whatever the nature of that link it is important to recognize that the quilt room is the source of the regeneration of the life of the domestic group and the act of weaving. This is evident since the nuptial bed is located in it and the woman head gives birth to cloth in reeling the first bobbin.

The internal social boundaries of the hearth are constituted through generational and gender differences, manifested most clearly in the domains of work and ritual. As a single unit the hearth also maintains its social boundaries *vis-à-vis* the kumba in that the former rests on what Uberoi (1971) has called 'conjugal blood'. This is seen in the way each hearth shares its property, housekeeping and commensality.

KUMBA

The kumba or the *kuf* (Ansari 1960) is the extended kinship group of the ego's father and mother. This is the smallest endogamous circle. It often refers to members living in a cluster of households, sometimes called *khandan* or *tabbar*. In Wajidpur, the 113 Ansari hearths comprise roughly twenty-three kumbas, with an average membership of thirty-two people per kumba (this includes migrants to Bhiwandi). The twenty-three kumbas are divided into seven Ansari bihaderis as shown in the schedule. On an average, each kumba is constitutive of five contiguous hearths and appears to the outside to be a single architectural structure. The four instances where different hearths of the same kumba were not contiguous with each other was due to the lack of physical space within the original kumba.

Relationships in the kumba are traced agnatically. In cases of a woman marrying outside the kumba, her father's hearth is not isolated from the others of the kumba. Instead, in a formal sense, the kumba plays host to the marriage feast. In times of marriage within the kumba, the bride's father's hearth is isolated from the other hearths of the kumba. This isolation is expressed not only through the absence of social contact, but also in the temporary abeyance of weaving relationships. From the point of view of relationships engendered via weaving, the bride's father's hearth is an affine to the rest of the kumba. It cannot solicit labour from other hearths until the bride has moved to her conjugal hearth. In effect, the bride's father's hearth is excluded from the collectivity composed by the kumba. This exclusion is expressed through the absence of weaving relationships. If we emphasize relationships cemented through quilt making we find the hearth opening its boundaries to the kumba in another sense. It is not isolated from other hearths. On the contrary, all kumba women congregate here to express their collectivity through an elaborate gift relationship.

As in quilt making, the kumba enters into the chulah during the circumcision ritual, but this time the entrants are circumcised males. The entry of kumba members for the ritual is significant since the novice gets access to his first group of socialization outside the household. As far as the novice is concerned, the significance of the presence of his male agnates during his ceremony lies in that he is now potentially a part of the productive life of the kumba. Here, too, a gift is made, one where the novice is gifted to his agnatic line by his MB. Read together with the gift made during quilt making it appears that the circumcision ritual begins what quilt making for the bride's trousseau ends — the full incorporation of the hearth into the kumba.

The kumba is an indispensable source of labour for weaving. Often individual hearths do not have the manpower essential for weaving, while others have excess people. Under such circumstances, depending upon the proximity of kin and affinal relationships with the hearth with excess workers, the labour-deficient hearth contracts from it. Barring wefting, this labour is utilized in any one or all of the other three stages of weaving. The return of services, either as gift, obligation or payment depends on the degree of kin relatedness.

If the ego has married his father's brother's daughter (hereafter FBD) services are provided on a reciprocal basis. Instead of a monetary transaction, the ego's hearth is obliged to provide its services on a similar occasion, such as building the FB's (father's brother's) hearth or marketing its produce in Barabanki town. The return of services is construed as a dyadic obligation. The same arrangement does not hold if the ego marries his classificatory FBD. In such cases the transaction is a monetary one with the understanding that the ego's hearth will provide its services later on the same terms and conditions. This agreement is valid in cases of marriage with the MBD (mother's brother's daughter) and the MZD (mother's sister's daughter) as well, as also the FZD (father's sister's daughter), provided the ego's FZ (father's sister) is not a member of the same kumba. If she is, then labour from her husband's hearth is never solicited, but always provided voluntarily by that hearth. The voluntary supply of labour extends outside the kumba to include the MB's hearth. Under no condition is the contract of labour financially transacted with this hearth.

While the supply of labour from within the kumba is solicited on the basis of kin ties, it is also dependent on the stages of weaving that labour is solicited for. It is instructive to see how weavers speak of labour solicited from hearths of the kumba. When kumba workers provide their labour for warping the term used to describe it is help (*madad*). This help, sought on the basis of kinship, is traced through previous obligations.

Adult kumba members do not like to provide their services for sizing and reeling since payment is subject to differing interpretations. Often the labour-contracting hearth evokes the kin ties between it and individual workers to defer payment, while workers demand cash since only hearths in dire financial need sell their labour within the kumba. Underlying this is the view of kumba members, especially male adults, that to sell labour to another for sizing and reeling is to lose honour.

Within the kumba, work for the first three stages is done by all the hearths' children who fall between the ages of six to twelve years. Male children enter this labour pool after their circumcision. These children are neither paid for their work, nor are their services determined through kin ties. A hearth is entitled to use the common pool of labour provided it had supplied its children to it on a former occasion. The utilization of this source of labour is a second area of potential conflict between different hearths because the number of children who constitute the labour pool is never enough to meet the labour requirements of all hearths of the kumba. It is, however, understood that children of a certain age-set will work for other hearths of the kumba. The phrase describing this obligation is the 'sacred earnings of children (*bachchon ki rozi*)'.

The supply of labour from within the kumba is part of a wider nexus of social relationships which regulates access into particular hearths. In everyday life this access separates the inside from the outside of the hearth, and also one hearth from another. The inside/outside division within the hearth is effected both by ordering space and types of work. This is seen on the marriages of daughters and on deaths. In marriages all women of the kumba congregate in the marriage hearth, while cloth manufacture for the shroud orders the territoriality of the hearth by investing certain areas with ritual significance, as well as correlating certain categories of people with determinate places and times. None of these categories belongs to the hearth for which this cloth is made. The MB or WB (wife's brother) cures yarn in the platform of the mardana from sunset to sunrise. The FZ or BZ (brother's sister), the wife of the agnatic head of the kumba and other married women who belong to the highest generation of the kumba, collectively reel bobbins in the zanana from sunrise to sunset. The FBS (father's brother's son) or BS (brother's son) makes warp members in the karkhana from sunset to sunrise. Finally, the agnatic head, who is also the kumba head, forms the weft members of cloth from sunrise to sunset. Terminologically, each of these people is referred to as mama, *phupphi, bhatija* and *bade abba*. The smallest family unit of the Ansaris must include these four categories.[1]

In this respect, the circumcision ritual shares a similarity with manufacture of cloth for the shroud. Here, the courtyard of the zanana is invested with sacredness and the nuptial bed is seen as the source of the regeneration of the domestic group, evident in that the removed part of the prepuce is buried under it. While the courtyard acquires significance through ritual performance, the nuptial bed is seen as the potential completion of such performance.

There is, however, an inner boundary between the two rituals. If through cloth manufacture for the shroud the work of weaving

---

[1] In the death ritual each stage of the weaving process is performed successively, with no overlap between them. Temporally, a relationship is established between the light and dark in that each stage is ordered by night and day. Night time should be understood as the period from the last call till the first call, or from the waxing and waning of the moon. The idea of light and dark is built into the ritual calendar. Ansaris say that the calendar itself is divided into a light and dark part. This notion of time will be explored in the third chapter. Suffice it to say that the weavers were known as the nurbaf (weavers of white light). The entire range of meanings implied by the term will be explored in the third and fourth chapters.

is constituted into four categories of workers, separated on the basis of gender and generation, the circumcision ritual shows how male and female are combined. This combination is inscribed on the novice's body by the mother and the barber in their respective ways. The mother, in preparing the body of her child, gives to him the gift of milk. This gift is feminine in that it is thought to be nurturing. It counterbalances the blood the child has inherited from his father. The barber, in removing the prepuce, constitutes the body of the novice as *hamdami* (of one breath). This hamdani is of the male and female, of passion (*shauq*) and its reception. This hamdami is inscribed on the novice's body through the ritual wound. In either case, of the mother and barber, the combination of male and female enables the novice to enter the life of the domestic group and the community of Islam. We might say that if the manufacture of kafan cloth separates the family into its constituent units, the circumcision ritual combines these units. In both cases, of separation and combination, the kumba plays a pivotal role since its entry into the domestic domain makes possible separation and combination. The domestic group must be situated between these two rituals: the circumcision of its male offspring and the death of its individual members.

BIHADERI

In Wajidpur the twenty-three Ansari bihaderis may be subdivided into two groups: the *Bharaichiya, Purabiya, Faizabadi* and *Multani* constitute the first group, while the *Shaikh, Madari* and *Muhammadi* compose the second. The first group associates the area from which the bihaderi migrated with the name of the bihaderi. The second group, in the estimation of Ansaris, comprises of the original inhabitants of Barabanki.

The term bihaderi denotes a circle of classificatory relatives, usually living outside the kumba, whom the ego is allowed to marry. A child's bihaderi includes the mother's siblings. These maternal relatives, especially the MB, play an important role in determining marriage preferences. The MB is responsible for maintaining an extensive gift relationship with his ZH's (sister's husband's) hearth. Classificatory relatives supply most of the contracted labour for weaving. Also, within this circle most Ansari weddings occur.

If the entire kumba is short of labour the circle of classificatory relatives, living in separate kumbas, is approached for providing labour. The contract is financially transacted, though the obligation to provide one's labour is informed by kin ties. This fusing of contract and obligation is most apparent in the case of a woman whose natal hearth is not located outside the kumba. With the exception of her elder brother and father all her relatives sell their labour when asked. This labour is spoken of as 'sacred earnings (*rozi ki kamai*)' or 'earnings that have been sanctified (*halal ki kamai*)'. Selling labour for non-weaving activities, such as loading trucks with animal fodder, is spoken of as earning to satisfy one's food requirements (*pet ki kamai*).

Spatially, the bihaderi does not occupy a distinct territory. There is, however, a spatial distinction made by weavers when they talk of differences between the kumba and the bihaderi. This is most marked in the case of the FZ and the MB when labour for weaving is contracted. Together, these two cases indicate how proximity and distance are ordered.

If the FZ is not married within the kumba, her husband's hearth's services are readily sought, as are those of his immediate agnates. If married within, this hearth is among the last to be asked for such help. In the latter instance this help is provided voluntarily until the ZH's eldest nephew or niece, living within the kumba, is married. After the marriage of her eldest nephew the FZ ritually distances herself from her brother's hearth through a ceremony called the *dola ki rukwai* (stopping the bridal procession before it reaches its conjugal home).[2] The circumcision ritual

---

[2] The dola ki rukwai refers to an occasion when the groom is escorting his bride to her conjugal hearth to consummate the marriage. Outside the conjugal hearth he is stopped by his FZ and other women of the kumba. The bride is publicly examined for her physical characteristics. The FZ directs loud scatological remarks on the bride's physical attributes and advises the groom on how to perform on the nuptial night. The FZ also expresses sorrow for she will no longer enjoy the fruits of her brother's household since he will soon yield authority to his son. The reference here is to the property she exploits in her natal hearth during her lifetime but can never transmit to her children. Following customary law Muslim women share their father's property, but cannot transmit it. In return, the groom's father

offers a contrast to the contract of labour for the purposes of weaving. If the FZ is married outside the kumba then no male member is invited to the ritual. If married within, it is obligatory for the hearth where the ritual is enacted to extend an invitation to the FZH's (father's sister's husband's) hearth. Taking both weaving and the ritual together it can be said (in the case of the FZ) that what the ritual combines (in this case the agnates, including the FZ), weaving separates (brothers from sisters).

If the MB is a member of ego's kumba, and provided ego is not married to the MBD, the attempt is to maintain a formal relationship with the MB's hearth. The MB occupies a position of ritual significance, but enjoys little contact with ego in everyday life. This is evident in the ties engendered through weaving. In daily life, ego's hearth does not lend weaving implements, such as the spinning-wheel or bobbins, to the MB's hearth, though on ritual occasions the latter is involved in sizing.

If the MB is not a member of ego's kumba then the MB-ZS relationship is a cordial, informal one, but also the most political. The MB is often mobilized to support his ZS (sister's son) in the latter's quarrels within his kumba. He is also a key figure in communicating his nephew's choice of spouse to his parents. In terms of weaving, the nephew is informally tutored on the details of the craft. Often the ZS spends days in his MB's hearth learning how to weave, provided the MB's daughters are not of marriageable age. It is as if this hearth provides a second home to him. If the ego is unable to establish his own dwelling within his father's kumba his next preference is his MB's kumba.

The significance of the MB from the point of the circumcision ritual is much the same as it is in weaving: the MB, irrespective of his location, has an important ritual role to play in his ZH's hearth. In this case he formally gifts the novice to his (the novice's) father's agnatic line.

Together, the case of the ZH and the MB show the degree of accessibility to the hearth. The MB's hearth must be situated outside the kumba for him to enjoy full accessibility into his ZH's hearth. The case of the FZ is complicated. If situated outside her natal kumba, the services of her husband's hearth are readily

---

makes a gift of money to his sister. After the ceremony, the FZ gradually reduces contact with her brother's hearth.

sought. If situated within, these services are never solicited but always provided without request. The circumcision ritual shows her significance from the opposite point: situated outside her kumba her male kin are not required to be present. If located within it is obligatory for them to attend the ritual. In the case of the MB proximity is premised upon territorial distancing while in that of the FZ, proximity is expressed more through ritual distancing than territorial separation.

Underlying the ordering of proximity and distance is a more general problem dealing with the nature of Muslim marriage patterns. Alliance is an important structural feature in cementing relationships of work in both ritual and everyday life, as well as connecting the hearth to the bihaderi.

Biraderi

Ideally marriage preferences for the Ansaris travel from the kumba to the bihaderi to the biraderi. The biraderi is the largest endogamous circle from which ego chooses a bride. Alavi (1972) says the term biraderi connotes a sliding semantic structure. The vertical axis represents the principle of descent while the horizontal axis telescopes the principle of fraternal solidarity. As one moves from the vertical to the horizontal one denotes a restricted kin group because of the condition of preferred patrilateral parallel-cousin marriage.

Commentators on this custom reiterate the Arab explanation that FBD marriage keeps property within the family. Property and blood are maintained and kept pure (Baer 1964; Granqvist 1931; Peters 1963). FBD marriage enhances the bond between the ego and his paternal uncle, since the latter becomes his father-in-law as well (Barth 1954); it facilitates the ecologically adaptive process of fission and fusion among the Bedouin by not building extensive affinal ties (Murphy and Kasdan 1959, 1967); it contributes to harmonious family ties since such ties already exist between the people involved (Khuri 1970). These arguments suggest that FBD marriage characterizes the affinal system as a whole, even when it is recognized that people do not marry either a real or classificatory FBD. Patai (1965: 347–8) says that in societies with this system of alliance, patrilineal descent is the only factor through which the ego relates to individuals and groups outside his own small world represented by his unilineal descent group.

This study, on the other hand, suggests that both alliance and descent are means by which a person relates to individuals and groups by augmenting and creating new relationships. A second way of linking the individual to the group is through work and ritual, both of which are enmeshed within kin ties. The assumed problem — FBD marriage does not lead to new alliances — is resolved by distinguishing the local line (the kumba) from the overarching descent line (the biraderi) by seeing how the latter traces its genealogy. If the local line is brought into focus, then in cases of FBD marriage alliance occurs between two hearths in the same kumba. Here residence is patrilocal and brothers live together. By the time their children are of second generation the brothers establish new and independent hearths and thus create new kin groupings, each with its own looms, the major property that they possess. At marriage the bride moves from her father's household to her conjugal hearth, leading to the creation of special affinal ties between two households that already enjoy an agnatic relationship.

The second problem — FBD marriage is preferred over marriage with other relatives — is disproved by considering the mechanisms and strategies employed in negotiating a proposed marriage relationship (Das 1973: 30–45). In this study such strategies are determined in part through relationships engendered by weaving. The question to be asked is not what is the nature and extent of FBD marriage, but under what rules and conditions does an Ansari contract marriage? This will be discussed later in the chapter.

In its widest conflation biraderi as a descent group encompasses all those between whom actual lines of common descent are traced in the paternal line, irrespective of generations that have elapsed. Thus, a determinate past is appropriated by the biraderi, as is apparent when Ansaris say that their genealogy may be traced to Ayub Ansari who lived in Medina at the time of the Prophet. From Ayub Ansari they take the name of their community. As weavers, they trace their genealogy to Sis Ali Salam, the youngest son of Adam, who was instrumental in fostering the sacred work of weaving in the world. This latter story is the subject of the fourth chapter. For the present, this tracing of descent shows that the biraderi is indefinite in size. On the other hand, written genealogies are not maintained by the Ansaris. Consequently, without these records the boundary of the biraderi is determined by the spread of existing and remembered kin ties. The limits of recognition of a biraderi imply that its genealogical depth is relatively shallow, and that the principle of descent has been arrived at by the actual, as opposed to the fictitious, demarcation of the descent line.

A biraderi connotes more than a descent group. Among contemporaries it cements lateral kin and affinal ties, best exemplified in the obligations initiated through weaving. Thus, the biraderi emphasizes both the solidarity of dyadic relationships and the spatial spread of the households of a biraderi of recognition. In its narrowest, most restricted sense the term signifies territorial specificity by distinguishing between members who are dispersed and those who are residents of a specific cluster of villages. Here, the biraderi may be considered a bihaderi or a kumba. In the context of this study Alavi's formulation that the movement from the horizontal to the vertical axis leads to a restricted kin group is a partial case, limited to marriage within the kumba, since more often than not the ego marries a classificatory relative living outside the kumba. Second, the kumba renegotiates the idea of siblingship and adulthood through the work of weaving, and male and female through the ritual of circumcision. Until the time that agnates inhabit the same hearth it is understood that their participation in weaving is not contracted. The contract on this occasion is *haram* (sacrilege, profanity). The partitioning of the hearth occurs both with the inheritance of the loom and the explicit understanding that work on a loom that is not one's own is labour (*naukari*).

TERMINOLOGY, MARRIAGE AND FISSION IN THE DOMESTIC GROUP

The above is a brief sketch of the basic kinship institutions of the Ansaris. Their working is seen in the use of kinship terminology in the circumcision ritual, the determination of marriages and the transmission of the loom, leading eventually to fission in the domestic group.

Terminology

In contrast to the Hindu system of marriage, Muslim canonical law does not prohibit cousin marriage. Among the Ansaris cousin marriage is considered the preferred norm. This norm complicates the terminology used to designate kin and affinal ties, which to the outsider often blur the boundaries between relationships through blood and those through marriage. For instance, among Hindus the term *bhai* designates all classificatory brothers, i.e. those cousins with whom marriage is prohibited. The Ansaris use bhai in the same way, but with them marriage with bhais is permissible and practised. Second, because of the practice of cousin marriage, the ego may share with his alter both a kin and affinal tie. Referential terminology does not deal with this complexity. Terms of address taken with those of reference show how Ansaris order their kinship universe.

In its referential system Muslim kinship terminology is similar to North Indian Hindu kinship terminology based upon the patrilineal family. As with Hindu kinship terminology, that of the Ansaris reflects differences of generation, age, gender and paternal, maternal and collateral relationships. These differences are designated according to the degree of kinship (by consanguinity or affinity), by age and sex, and by simultaneously expressing differences of behaviour between members of the family, the conjugal couple and the kumba depending upon the individual's place in one of the categories. Only in the ego's own generation is there a merging of distinctive kin types into a single category within the immediate kindred. The ego's primary reference terms classify more distant relatives with whom a relationship either through marriage or descent may be traced.

As far as address terminology is concerned a limited number of terms are utilized, these being variations of consanguineal reference terms. At the outset four general principles must be kept in mind while using address terms.

(1) A senior person is always addressed by a kin term, never by the first name. In the case of more than one senior of the same sex occupying the same genealogical category, address terms reflect differences in age. For example, for two older brothers *jan* and *sahib* are suffixed to the address term. Juniors, too, have address terms though it is not obligatory to use them. Here, the

use of an address term is often a device distancing the ego from the addressee. During the weaving cycle the ego uses address terms to designate those workers with whom he shares a non-monetary relationship, and reference terms for those with whom his relationship is contractual. In the circumcision ritual discussed below we see that the use of such nomenclature often expresses political relationships between two hearths (see Chart I).

Chart I shows seven members of the novice's generation, eight of the first ascending generation and one of the second ascending generation present in Sadiq Ali's son's ceremony. On the day of the operation each member, excluding the boy's parents and the barber, was introduced to the novice by a genealogically appropriate term. In his generation males were designated as bhai. Henceforth he would be urged to spend part of his day with members of this group.

In the first ascending generation a combination of reference and address terms designated each member. The difference between terms of address and reference reflect both kin relatedness and grievances with particular households. In the present instance, the head of B^4 married to a woman of C^5 made the gift of the novice to the barber. Hearths A^2 and C^5 had developed an acrimonious relationship. Sadiq Ali believed that B^4 and C^5 were in collusion with each other because of their close affinal relationship. He, however, felt he could counter their machinations since hearth A^2 had a steady supply of labour in the lowest generation (see Chart III). Consequently, through his wife's brother he introduced the head of B^4 to the novice as his *taya*, not *dada* or bade abba as should have been the case. The insult was doubly compounded since Sadiq Ali's elder brother was introduced as bade abba. Often, such terms of designation become instruments by which cliques are created.

(2) Seniority within the kumba is of two kinds: age and relationship. The former are not referred to by their first names, even if located in the ego's first generation. When located in a generation lower than ego's, address terms denoting the closest consanguineal relationship are used. Such instances are frequent. The following example is illustrative. (Chart II).

In this kumba Ishtiaq Ansari and Aslam Ahmad live in separate but contiguous hearths A^1 and B^2 respectively. Mumtaz is a member of the same kumba who, before marriage, lived

CHART I: TERMS OF ADDRESS IN THE CIRCUMCISION CEREMONY OF SADIQ ALI'S SON

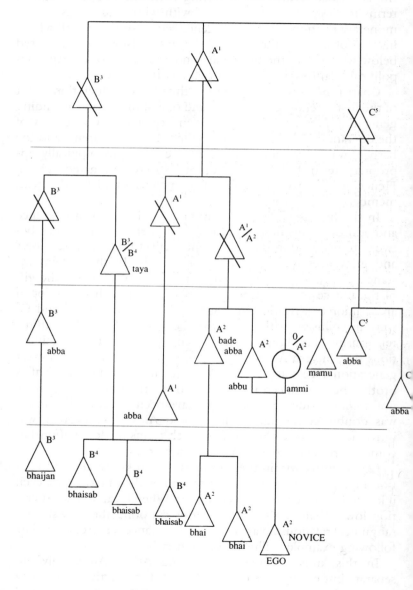

Household, Kin, Work and Ritual • 55

CHART II: THE USE OF KIN TERMS OF DESIGNATION

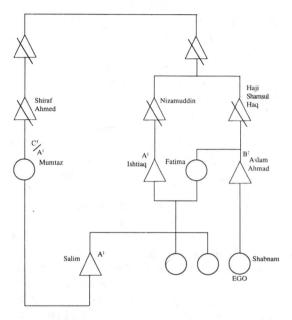

in hearth C^3 with her youngest brother and mother. After marriage Mumtaz moved to Shabnam's FZS's (father's sister's son's) hearth A^1. This man is Shabnam's classificatory bhai, while Mumtaz is Shabnam's classificatory phupphi, though she is younger than Shabnam. After Mumtaz's marriage with Salim, Shabnam continued to address her as *chhoti phupphi*, though in referential terms this phupphi is also her *bhawaj*, brother's wife (BW). If Mumtaz had come in to Shabnam's father's hearth — if she had married Shabnam's brother — she would be addressed as *bahen*, not chhoti phupphi or dulhan or *bhabhi*. Interestingly, in the context of terms of reference, Shabnam's FZ, Fatima, is also through an affinal relationship, her classificatory *chachi*, though Shabnam continues to address her as phupphi. Fatima's offspring, however, are referred to as *chacha zad bhai/bahen*, and not as *phupphi zad bhai/bahen*.

(3) Consanguineal members of the same generation and their spouses who are older than ego are addressed by their genealogically appropriate term. In cases of marriage between the ego's

relatives, the kin terms that address them prior to marriage are used by the ego, provided they are not members of his hearth. When a bride enters his hearth the term *baji*, if she is older, and *dulhan*, if younger, is used to address her. Dulhan is replaced by the woman's name or *apiya* after the circumcision of her first male child and/or when she begins to assist her mother-in-law in reeling. This also corresponds to her integration in the zanana. For older kumba members of the same generation a suffix is often added to their genealogically appropriate kin term. Depending on the context, the suffix connotes respect in one instance and mockery in another. This ability to play with kin terms is restricted to relatives living within the bihaderi but outside the kumba. With the exception of the MB's household the ego uses a limited number of terms that usually designate the generation of the addressee: mama, bhai and *beta*. If the age difference between ego and addressee is negligible then the former addresses the latter by his name. This is most pronounced when bihaderi members supply their service for weaving.

(4) To designate one's affinal relatives ego generally uses terms employed by the spouse. Spouses do not address each other and avoid all direct communication in public. With his mama living outside the kumba the ego often enters into a playful terminological game by establishing a kinship relationship which reduces the MB's generation to his own.

The above rules are most visible in an Ansari hearth of three generations. This type of hearth is composed of parents and, in the second generation, their children, married and unmarried, and the married children's spouses. The third generation comprises the grandchildren of the first generation. These children have not yet reached the age of marriage: girls are not in *parda* and do not stitch quilts, while boys do not operate the loom. A two generation hearth includes a man who has partitioned from his father's hearth, his wife and their unmarried children.

In both these types of hearths children do not directly address their real and classificatory fathers (reference term *abba*, address term *abbu*). Their relationship with their real and classificatory mothers is less formal (reference *amma*, address *ammi*). Elder siblings living in the same hearth interact freely with their younger sibs and often call them by their first names or nicknames. The use of names for cross-sex sibs is infrequent between those who live in different hearths. Here the elder brother addresses his younger sister as *bajiya/apiya* if she lives in the same kumba, but as *baji/appa* if she lives in a different one. In the case of brothers living in two different hearths marriage between their children does not influence the terminological relationship between them though it alters their weaving relationship.

Younger sibs must show respect to their elder sibs living in the same hearth and refer to and address them as brother and sister. As a rule this includes older affinal relatives. Between brothers living in different hearths the terminological relationship may be modified if the younger brother is richer than the older, often signifying that his is a two loom hearth. Their weaving relationship, however, remains the same. If the need arises they contract labour from each other in the same way as it is contracted with the FBD's natal hearth when one has married the FBD.

The sibs' relationship with their FB is characterized by restraint, especially if he is older than their father. In case the father is dead, his eldest surviving brother represents the hearth on all formal occasions, such as leading its male members for the Friday prayers and initiating marriage negotiations for his nephews and nieces. Terminologically, the change is reflected in address terms — from chacha or taya to bade abba. Weaving arrangements between such hearths are never contractual but always reciprocal.

Within the hearth and kumba address terms signify a determined mode of behaviour and reveal how a new entrant is to behave. This is most evident in the case of a bride, who until she renegotiates her status in her conjugal hearth, is referred to as dulhan by those who occupy the ascending generation and those who are older than her. Till the time she has her first child the dulhan does not maintain eye contact with the elder males and females of the kumba and the hearth. In the hearth the first person she shows her face to is her HyB (husband's younger brother, *devar*), and the last person, her HF (husband's father or *sasur*). After her first child her HM (husband's mother, *sas*) and HF call her daughter (*beti*). If she is a member of the same kumba then, with the exception of her natal hearth, all kumba members who are older than her call her dulhan, but with outsiders to the kumba, refer to her as beti. The term dulhan is dropped and beti substituted when the bride participates in her first son's circumcision ceremony and/or assists her HM in quilt making and thus moves into the zanana of her conjugal hearth. She also begins reeling.

By itself the kinship terminology of the Ansaris does not show how the group perpetuates itself. While it marks the differences between insiders and outsiders it does not explain how such differences arise. These distinctions are explained by the processes of marriage and fission in the domestic group.

Marriage and Fission

Broadly, fission occurs because of marriage. Three general principles regulate marriage. The Ansari marriage system is based upon preferred cousin marriage with one limitation: in the same generation a hearth does not contract marriage with another hearth more than once. This means that two brothers will not take wives from the same hearth. Cross-sex marriage between two hearths is not approved of. Marriage is confined to the biraderi and an alliance outside it leads to social ostracism by which the guilty couple is expelled from the biraderi. I will explain the dynamics of fission by a case study. (See Chart III).

In January 1986 Sadiq Ali's kumba, divided into five hearths, had a living membership of thirty-six people. I have shown the division into hearths in the following way.

Hearth A^1 — Ego's(Sadiq Ali) FF. Two looms, two generations.
Hearth A^2 — Ego's father. Formed from the partitioning of A^1 Two looms, one of which is hereditary. Two generations.
Hearth B^3 — Ego's FF2eBS. One loom, three generations.
Hearth B^4 — Ego's FF2eByS. Formed from the partitioning of B^3. Two looms, one of which is hereditary. Three generations.
Hearth C^5 — Ego's FFyB. One loom, three generations.

In the chart I have used the following abbreviations:

A^1/OTHER, etc. — Member of a hearth who has permanently left the kumba.

OTHER/A^1, etc. — A member entering the hearth from outside the kumba. Except in the case of the

Chart III: Sadiq Ali's Agnatic Line

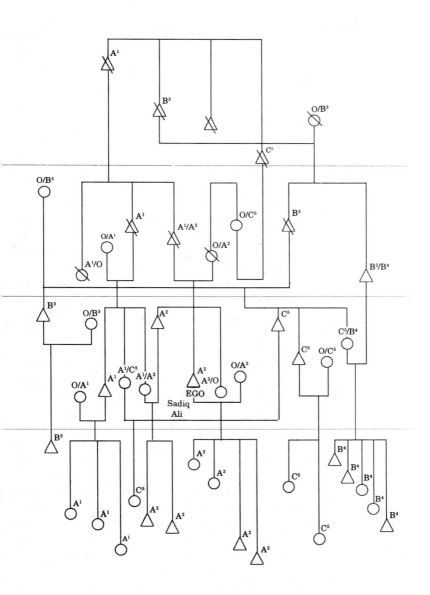

	ghar jamai it is always women who either permanently leave or enter the kumba.
A^1/A^2, etc. —	In the case of males these symbols denote that A^2 has formed from the partitioning of A^1. In the case of women it denotes movements from A^1 to A^2 through marriage or adoption. Movement here is within the kumba.

In Sadiq Ali's kumba the fissioning of hearths follows a pattern where the youngest brother establishes his own dwelling. When his brother's children reach marriageable age then Sadiq Ali, the ego, will partition from his parents' hearth. This process has begun. Sadiq Ali's elder brother's eldest daughter has reached the age of marriage. Sadiq Ali looks forward to the establishment of his own dwelling since it will give him the hereditary loom. His brother Ashraf has mixed feelings about this partition. Ashraf had recently acquired a new loom after taking a loan from a village bank and is hard pressed to meet payment deadlines. Fission would mean that Ashraf is denied the surplus labour hearth A^2 enjoys. Having to contract labour would eat into his meagre profits. Also, fission would involve not only the movement of the hereditary loom and the division of the movable property of A^2, but also financing the construction of the new dwelling.

The fissioning of B^3 and the consequent establishment of B^4 has followed the familiar pattern of the elder brother retaining control of the hearth. The problem, as the head of B^4 sees it, will arise when B^4 is to be partitioned. Two new hearths will have to be established and the space for them marked out in advance. In finding a space for his sons to settle down the head of B^4 is constrained by various factors. Occasionally, because of the spatial limitations to the boundaries of the kumba male members are forced to migrate outside the kumba to other villages or urban areas. The first preference is to establish settlements in their MB's kumba. If space is unavailable here the male usually lives within the environs of the village of his father's kumba, or in a contiguous village. If within the village but outside the kumba, he is treated as a kumba member. If he lives outside the village then frequent contact is necessarily limited, especially among women. The hereditary loom will never move outside the kumba. This territorial rootedness of the kumba makes the occurrence of the ghar jamai more frequent than it is assumed to be. The head of B^4 says that one of his sons will have to migrate outside the kumba. Since this head is married to his FBD, his sons are already members of their MB's kumba. Nor can the son who is separating settle in his MFBS's (mother's father's brother's son's) kumba, it being the same settlement. The head of B^4 is forced to look for space in the kumba of the brother of the woman head of hearth C^5, the classificatory WB (*sala*), whom he addresses as brother (bhai).

From the above two cases we see that the kumba maintains its territorial boundaries through the transmission of the loom. Because the hereditary loom must remain within the kumba, the attempt of the fissioning hearth is to find a dwelling within the territory of the patrilineage. Also, fission leads to a realignment of social relationships between the newly built dwelling and the rest of the kumba. This finds its clearest expression in relations engendered through weaving. When a newly built dwelling, inhabited by two generations, solicits labour from the kumba it must show evidence that it is financially independent of its parent hearth. This is done by meeting production requirements. Simultaneously, this dwelling shows signs of participating in the communal life of the kumba, achieved primarily by making its children available to the common pool of labour. Finally, this is a significant period for the other hearths to express resentment and conflict with the newly emerging hearth by explicitly demanding that the supply of labour be financially transacted. In cases of serious conflict the status of a two generation dwelling may be ignored altogether by refusing to supply labour to it.

Sadiq Ali mentions that though his household has the most number of workers living in it, it is not yet a fully formed hearth. A chulah, he says, is one which has three generations living in it and a hereditary loom. A loom becomes hereditary in two ways: it is either transmitted across generations or cloth for the shroud has been woven on it. While A^2 has a hereditary loom, it is not composed of three generations. Also, while its financial independence from its parent hearth A^1 is taken for granted, Sadiq Ali says he has to continue supplying his children to the common pool of labour despite needing workers himself.

A^2 shares a long rivalry with C^5 which can be traced to the ego's father's generation. Sadiq Ali says that his mother's eldest sister, the present woman head of C^5, has been callous to his FBD who is married into C^5. As a result of ill-treatment she is unable to bear any more children. Her husband, the present head of C^5, had twice sent her back to her natal hearth because he wanted a male offspring. It was only Sadiq Ali's brother, married to the younger sister of this woman, who persuaded the husband to take back his wife. Hearth C^5 took offense at what it considered unnecessary interference in its affairs. It began to demand monetary payment for the labour it provided to A^2, and refused to utilize the services that A^2 was obliged to provide for weaving. C^5 acted in the way it did because A^2 is a two generation hearth.

If fission is an occasion when conflicts are expressed, it is also important to recognize that the cohesion of the kumba is expressed through the horizontal fission of hearths in the male line. Here the general principle of fission is explained through marriage: when in the same hearth the third descending generation reaches the age of marriage, the second descending generation partitions along the male line. This indicates that a particular generation has reached the age of marriage, and also reworks siblingship and adulthood. Formal rules do not commemorate this re-working, though there are markers in the lives of individuals.

For women, the veil separates as well as unites. Ansari women in Wajidpur do not practice parda in the way of upper-caste Muslim women. They do not wear the *burqa* within their natal and conjugal kumba. Within the kumba they cover their faces in the presence of certain relatives. If unmarried, parda is maintained in the presence of those they can marry. In the lateral spread this includes the FBS (*chacha zad bhai*), the MBS (*mama zad bhai* or mother's brother's son), the FZS (*phupphi zad bhai*) and all other classificatory brothers. Within the hearth, the woman who goes into parda is spatially relocated. Most of her time not spent in weaving is spent in the zanana making quilts. Her interaction with males is drastically reduced. She is placed under the watchful eye of her elder sister or mother, who tutors her in her conduct with males. While she continues to cure yarn for weaving, most of this occurs in the zanana. It is acknowledged that a woman's work in her natal hearth can never be taken for granted, for she is a transient to be eventually incorporated into her conjugal hearth. Parda thus involves the closure of former groups of acquaintances and the opening and redefining of new and previous associations.

The closure and redefinition of social relations engendered by parda is seen in the behaviour of the bride (dulhan) in her conjugal hearth, especially in the work she does for weaving. The bride is required to supervise the sizing of yarn, executed not in the zanana, but in a compound contiguous to the mardana. While she does not wear the burqa in this supervision, she does not directly communicate with any of her conjugal hearth members, except children. She must keep her eyes lowered when addressed by elders and refrain from voicing any opinion in public. The only people she conducts a conversation with are women of her lateral spread, and, privately, her husband. Until her first child the bride does not work in the zanana. After the circumcision of her first son she begins to cure yarn in the zanana. Also, she enters into a joking relationship with her HyB (devar) and begins to communicate with the younger male members of the kumba. The removal of the veil corresponds to her integration into the zanana of her conjugal hearth. Terminologically she is addressed as daughter (beti), not dulhan, by the ascending generation, and as sister (apiya), or by her first name, by members of her own generation. While for the daughter parda is a forerunner of fission in that it implies her marriage and the birth of children, in the case of the bride it shows how spaces are marked through work.

Among males the marker announcing fission is a realignment affected in the division of labour of weaving. When the male reaches marriageable age, and if he is not the head of his hearth, he begins to participate in wefting. The movement is from providing labour to the weaving process, such as fixing the heddle frame on to the loom, ensuring the warp is straight, to the operation of the loom itself. The hearth head delegates responsibility to his eldest son, finally culminating in marketing the cloth in town.

If weaving crystallizes the identity of the male as the head of the hearth, it may be recalled that the genesis of this mark is inscribed on his body through circumcision. In a later chapter I will show that through the ceremony the novice's biography is plotted in the domestic group and the kumba. Potentially, such plotting makes available to him his position as father, taya, chacha, mama, on the one hand, and his reproductive and productive life, on the other. In this sense the circumcision ritual establishes a relationship between the novice and his first ascending generation by which his biographical time is punctuated according to the kinship and weaving positions he will occupy.

In Sadiq Ali's kumba the daughter of his FBD (of hearth C^5) was, at the time of this study, in the process of going into parda. Informally, it was agreed that she would marry ego's classificatory brother's son (B^3 in the lowest generation). This man retained possession of his father's hearth B^3. Hearth C^5 would partition before the marriage. The older brother of hearth C^5 in Sadiq Ali's generation retained possession of his father's hearth. His younger brother had begun the establishment of his own dwelling within the kumba. His wife remained in her natal hearth (*maike*) during its construction since she was pregnant. The building of the new hearth was a collective effort. Besides every available member of hearth C^5, hearths A^1, A^2, B^3 and C^4 were required to supply labour in the construction of the new dwelling. Because hearths A^1 and A^2 were in conflict with C^5, at the time of supplying labour they demanded monetary remuneration. The object, Sadiq Ali explained, was not that hearths A^1 and A^2 wanted money, but that C^5 needed to recognize that the kumba was bound by brotherly ties (the term he used was *bhaiband*). Short of labour and money, the head of C^5 began to show concern at his wife's plight and went to the extent of saying he would adopt one of Sadiq Ali's daughters.

Terminologically, there was an interesting shift in the relationship between hearths C^5 and B^3. Once it was decided that the male member of B^3 in the lowest generation would marry the daughter of the head of hearth C^5, the latter traced its relationship with B^3 through the woman member of hearth A^1 married into C^5. By this the male member of B^3 in the lowest generation is the male head of C^5's classificatory sister's son (i.e. WFFBSSS or wife's father's father's brother's son's son's son). The male head of C^5 refers to him as *bhanja* (ZS or sister's son), though before this marriage was agreed upon, hearth C^5 traced its relationship with hearth B^3 through the father's line.

After the new dwelling is fully built the head of C^5 will formally gift his brother with the ancestral loom. This gift is called *nazr*, though it is part of the property the younger brother inherits. Nazr is always mediated by a third person, usually by a member of the community who has been on the pilgrimage to Mecca

(*Haji*). In turn, the younger brother provides a feast to all the male members of the kumba. This feast, called the food of the collective (*biraderi ka khana*), is supposed to signify the first produce of the ancestral loom by the inheritor.

In cases where the older brother establishes his own house, it is not the younger brother/s who provide financial aid in its construction, but the father. If he is not alive the eldest surviving brother of the father is obliged to underwrite part of the financial costs of the new dwelling. In such cases the FB performs the nazr. If no such male member is available the gift is made by the Haji to the inheritor with the classificatory FB present as witness. During nazr the Haji reads out the conversation that is said to have occurred between Jabra'il and Adam. I will discuss this conversation in the fourth chapter. What is important here is that the loom, when it is being transmitted, transcends the social boundaries of the hearth by reinforcing the blood bond between ego and his first ascending generation. It does this by establishing a continuity with the tradition of weaving and with the community's ancestors.

Work and the Hearth

From the preceding account it is evident that work and ritual are embedded in the kinship structure of the Ansaris. The meaning of, and attitude to, work is determined by the location of the individual weaver in the kinship structure of the group in its most general sense, and within the domestic cycle in its particular sense. First, both work and ritual establish a relation between men and women. In the circumcision ritual this relationship seeks to unite men and women, while in work it effects a separation between them. Second, a distinction is made between generations: males of the most senior generation work on the loom and provide verbal legitimacy to the ritual, while those of the most junior generation provide their services and are the subjects of the ritual. Third, work marks a difference between insiders and outsiders. Within the hearth this is found in the difference between consanguineal and affinal relatives. In turn, the hearth distinguishes itself from the kumba: members of the former work as a unit as far as weaving is concerned, while the kumba itself is differentiated on the basis of the FB's household and others. In the circumcision ritual the kumba functions as a unit, but under the formal authority of its agnatic head. The hearth and the kumba are together distinguished from the bihaderi. For the latter the basis for providing services is monetary in the case of weaving. The exception is the MB's household. However, the kumba and the bihaderi play a vital ritual role in the life of the hearth. This is seen both in the circumcision ceremony and when cloth for the shroud is fashioned. The biraderi, in tracing a genealogy to Ayub Ansari and the first man Adam, legitimates the noble birth of present day Ansaris and provides them with a profession that is sacred. I will consider the work of weaving in terms of its ability to categorize the space and time of the hearth.

This work is sacred not only in that cloth is made for the shroud, but also because the ritual act of weaving marks out the space and time of the household. This mark is seen in the distinction effected between the work domain of men and of women. Temporally, there is a generational ordering of the four stages of weaving. However, a further set of spatial contrasts flow through the loom. They deal simultaneously with the division of the body into an upper and lower half, and a division of the day into a light and dark part.

In his discussion of the Kabyle house Bourdieu (1977: 89–91) deals with a similar set of contrasts. He shows how notions about men and women, light and dark, inside and outside, are contrasts which derive from objective conditions: division by age, sex and position in the relations of production. These contrasts order both the conceptual and social domains of the Berber world. In other words, the organization of space is governed by the same set of contrasts that inform the practical and discursive knowledge of social actors. The meaning of a given spatial order is not merely the physical position of its constituent elements, but includes the activities of social actors and the economic and social conditions which inform these activities. This practice is itself rooted in a set of conceptual schemes represented in the order of space, but the actual meaning given to a spatial order depends on the nature of the activity concerned.

Bourdieu introduces the notion of 'universes of meaning' and 'universes of practice' (1977: 122–4) to explain how the organization of space has different meanings in different contexts. Meanings invoked in one context have the ability to refer to meanings in another, since the actual significance a given set of contrasts acquires in relation to a particular universe of practice is suffused with all the meanings these contrasts might be given in other fields of practice. In other words to interpret space is to ask what an element does in the spatial order. To understand what this element does is to analyse the physical movement of the actor in relation to his temporal and spatial activity in ordered space. This is what Bourdieu attempts. The spatial order, then, may be understood in the following way: (i) what is inscribed in the organization of space is not the actuality of past actions, but their meaning; as a corollary, individual events are superseded by the significance of what is done; (ii) since the spatial order both precedes and succeeds individual actors the significance of the order cannot be identified with particular individuals.

Given the above, I will show through an analysis of work performed on the loom how certain spatial contrasts, such as the vertical/horizontal, front/back, up/down, left/right, derive from divisions of generation and gender in the relations of production. In effect, I will show how work on the loom, most marked during ritual occasions, is related to the kinship structure of the Ansaris. The work shed shows that space has a definite territory assigned to it which is marked out in the act of weaving. To extrapolate from Bourdieu, space acquires meaning in the location of its practice. To interpret this space is to ask what the act of an actor does in a demarcated area. Thus far, I have only discussed the location of this practice. In what follows I will show how this practice constitutes its area.

The loom provides the following contrasts which have a spatial dimension. First, weavers distinguish between vertical (*tana* or *lambai*) and horizontal (*bana* or *chaudai*) members of yarn, or the warp and weft of cloth. This leads to a second distinction between the front (*age*), which is the yarn, and the back (*peech*), which is cloth as finished product. In the process of making cloth warp members are always considered to be at the front of the loom, while the weft is both at the front and back of the loom. The back is referred to as 'cloth at the back (*peech ka kapra*)', while the front is termed as 'yarn at the front (*samne* or *age ka sut*)'. Cloth at the back signifies the completion of one weaving cycle, while yarn at the front means that weft members remain to be fashioned.

A second set of spatial contrasts take as their point of reference the body of the weaver at work on the loom. The weaver differentiates between the upper (*onch*) and lower (*neech*) halves of his body. The upper half leads to a contrast between the left (*shuttle ka hathwa*) and the right (*muthia*) hand of the weaver. The upper half of the body, especially the hands, are in the estimation of weavers, responsible for fashioning the weft members of cloth, while the lower half of the body, especially the feet, fashion the warp members of cloth.

These contrasts, found in the *Mufidul Mu'minin*, divide the male weaver's body into an upper and lower half, with the loins of the weaver marking this division. The first du'a recited in everyday production refers to the girding of the weaver's loins. The second and third recited du'a are concerned with the movement of the shuttle by the right and left hands respectively, while the fourth du'a corresponds to the movement of the feet in the pit. In terms of the loom a similar distinction is made between the pit and the right hand. Any fault in the loom is located in terms of this broad classification.

For the loom there are two major sets of contrasts each with their own sub-set. To recapitulate, these contrasts are: the distinction between the vertical and horizontal members of cloth, leading to the distinction between the front and back of the loom. This is reflective of the difference between cloth as finished product and cloth in its raw form as yarn. Taken together, the two sets of contrasts deal with the distinction between the third and fourth stages of the weaving process.

Located within the loom there are two further contrasts, the second a derivative of the first. The first is a difference between the upper and lower halves of the weaver's body in relation to the loom. The upper half is again subdivided into the left and right hand of the weaver which corresponds to the left and right sides of the loom respectively.

The two basic sets of contrasts, each with their own sub-set are explained by the denotative and connotative content of the term nurbaf, which may be translated as 'capturer of white light' (Pandey 1983). Whenever they describe their story of origin Ansaris say that traditionally they were known as the nurbaf because they wove cloth of pastel shades by capturing the light of the sun (*suraj ki bunai*). The head weaver of a particular village was categoric in his assertion: 'deep colours are profane (*gehra rang haram hai*)'. Among weavers it is not only that a person wearing clothes with human representations is not allowed into a mosque, but that a male clothed in dark colours is prohibited from entering a place of worship.

While cloth is differentiated through colour, this distinction finds expression in the way the place of weaving is organized in terms of its activity. The territorial place of weaving within the hearth may be discerned by seeing which areas of the household are thought of as 'dark' places in relation to places that are considered 'light'. In this way the act constitutes its area, most evidently in the work performed in the work shed.

As mentioned earlier, the work shed is a room divided into two types of activities performed on two instruments of weaving. The warp beam, an instrument on which reeled bobbins are mounted, forms the vertical members of cloth. The loom forms the weft members over and below the vertical. With the operation of the loom yarn becomes cloth. In the weekly cycle such cloth is rolled and placed at the back of the weaver. Cloth for the shroud is not rolled but left in a heap. In the work shed the warp beam is placed at the front of the loom with the weaver facing it when at work. The warp beam is invariably located at the door of the work shed so that it receives the maximum light of the sun. When cloth is woven after sunset a lighted lantern is placed near the warp beam so that if warp members snap they can be tied together. This pragmatic explanation is partial since within the work shed the contrast between the front and back of the loom does not fit with the above explanation. The front/back dichotomy is related to the distinction between cloth and yarn. The vertical members of cloth receive more light than the horizontal and the front more light than the back.

The second set of contrasts distinguish between the upper and lower halves of the loom and of the weaver's body. The upper half further distinguishes between the left and right of both the loom and the body. In comparison with the lower half, the upper half of the loom is more illumined. The lower half refers to the pit in which the weaver's feet operate two pedals that are synchronized with the movement of the shuttle from the left to the right. The lower half is not divided into a left and right because the feet perform the same function. Within the upper half, the left side of the loom is the side of the shuttle, while the right is the side of the grip. The right side of the loom is thought to receive more light than the left since it is through the grip that the weaver grasps the light of the vertical members of yarn. This is the main characteristic of the grip. The diagrammatic structure of contrasts is given below.

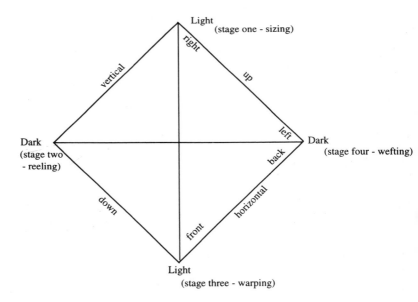

Within the remaining areas of the hearth one or other of the contrasts embodied in the work shed is expressed. In the weekly cycle, sizing and warping are associated with light colours, while reeling and wefting have the characteristics of darkness. In this cycle yarn is cured either in the compound of the zanana or of the mardana. Sizing is executed in daylight since yarn must be dried in the sun. In contrast to both sizing and warping, reeling occurs in the shaded part of the hearth. Ansaris say that the first bobbin is reeled in the quilt room, never in the mardana. Bobbins are reeled only after the woman head of the hearth has supplied the first bobbin. Before she supplies this bobbin she blows on it so that it is excised of blemishes. In the weekly cycle the two weaving stages associated with the darker areas of the hearth involve activities that are performed during daylight. In the case of reeling the act begins by about mid-day, while in wefting it starts immediately after the morning call. Daylight is timed from the first call. In making cloth for the shroud this order changes.

In making cloth for the shroud sizing occurs in the mardana of the hearth after sunset. Reeling occurs before sunset of the subsequent day in the quilt room, which is the darkest part of the hearth. Bobbins are attached after sunset. Finally, cloth is woven after the morning call of the next day. In making this cloth the spatial structure occasioned by weaving is retained: the division between light and dark areas does not change. What is magnified is the contrast between the inside and outside — of the hearth and of the personnel who work on the weaving stages. In addition to hearth members, the inside of the hearth is constituted by those who work on the four stages of weaving. Each of these members is thought to represent certain qualities. In making cloth for the shroud they impart these qualities to the corpse. This is explored in the third chapter. The person who cures yarn — the MB for the middle generation of a three generation hearth — may be considered an outsider to the extent that he does not belong to the same agnatic line of the hearth. In fact, through his participation as wife-giver, it is recognized that the agnatic line must look outside its restricted group to perpetuate itself. This notion of the outside is given spatial expression in that the MB must be a member who does not belong to the same kumba. In this way, too, the status of the woman head of the hearth is constituted through marriage. I was told that as far as marriages are concerned a spouse is sought, first, among consanguines of people who work on the shroud. In effect it points to the entry of the bihaderi in the hearth.

The entry into the hearth of relatives living outside the kumba shows that it is constituted, at least partially, through marriage. The entry also gives to the ego his basic terms of address. Apart from the terms he uses to address hearth members, the terms are bade abba for taya (FeB), *ammi jan* for phupphi (FZ), mama for *mamun* (MB). For members of a generation higher than him these terms connote formality and proximity. For members of the same generation they imply a brotherly relationship, but always tempered by distance.

In making cloth for the shroud the spatial division of the hearth into male and female zones is reiterated. In sizing, the yarn moves, as it does in the weekly cycle, from the outside to the inside of the hearth. But here, this movement does not imply the renegotiation of anyone's status. One reason why yarn is dried inside the hearth is because on such occasions no woman of the hearth steps outside the zanana. More importantly whenever a single man cures yarn on the platform of the zanana it signifies that yarn is being prepared to make cloth for the shroud and that the hearth is closing its boundaries to the outside. This closure has its spatial correlate in that the remaining three stages of weaving occur in the inside of the hearth — either in the zanana or the work shed.

The contrast with the circumcision ritual is instructive for if weaving structures the space of the hearth from the vantage point of the mardana and the karkhana in particular, the circumcision ritual structures it from that of the zanana. Like the mardana the zanana also provides a set of spatial polarities during the ritual, but not as clearly marked as in the case of weaving. First, through the courtyard and the quilt room, the zanana establishes a contrast between the sparseness of the former and the fullness of the quilt room. Following from this we find a directional orientation in the courtyard in that the ritual is enacted with the actors facing the western wall, which is especially cleared of all impedimenta. It resembles the western wall of the mosque. In the quilt room, on the other hand, the orientation is organized in terms of the underground (the prepuce is buried under the nuptial bed) and the overground, more appropriately, between the surface and the depth.

In turn, the contrast between the surface and depth and the significance of facing Mecca are made visible on the body of the novice. We find here a triadic classification of his body: depth, surface and height. The depth of the body is thought to be composed of dangerous secretions, which, to be excised, must be brought to the surface. The operation is the act of uncovering the depth and in the process imprinting the boy's body with a regime of signs. These signs show how the body acquires a socially recognized materiality. Organized on the surface of the body, they are legitimated during the ritual by whispering the azan and the boy's formal name in his right ear. Thus, the body is composed of a dangerous and polluting depth, a surface organized in terms of signs and a height where the novice is impregnated with the word of God.

The territoriality of the hearth both marks out and is marked by the act of weaving and the circumcision ritual. That is to say, both weaving and the ritual create their own spatial thresholds as much as they are circumscribed by their locatedness. Place here is not location in geometric space. If it was then place would be nothing more than the location of physical bodies and their properties. As I have used it, place is an item of some equipment, both material and symbolic, that belongs to a context determined by the equipment necessary for the work of weaving to be done. Before these places are assigned a context, however, a region has to be discovered in which they will appear in their necessary connection. Depending upon the type of cloth that is woven, for the market or for the corpse, the region may be the hearth or the lineage. In either case, the connection with the kinship structure of the Ansaris is basic. This base shows the type of personnel implicated in the work of weaving and of the ritual. It does not, however, explain the technical details of the craft of weaving or of the ritual and the way both constitute their world. I will turn now to a consideration of the way work is constituted by examining the rules that frame the endogenous tradition of weaving.

Chapter Three

The Semiotics of Weaving

INTRODUCTION

In the endogenous tradition of weaving time and space are encoded in a way that their constituent units are sanctified in the act of weaving. The common thread running through the encoding of space and time is the location and spatialization of the body of the weaver in relation to the instruments of work. Thus, in the production of cloth the cosmological ideas informing the world of weavers are expressed.

In focusing on the production of cloth, the Ansaris privilege two areas: colour and design. This is seen most dramatically when cloth for the shroud is woven. It is not only that colours and designs are represented upon the cloth for the shroud, but also that the different parts of the body of a dead man are marked through these representations. Each of the colours and designs is identified with a specific bodily trait and a particular instrument of weaving. Finally, in making such cloth, workers typify the qualities signified by these colours and designs.

While the cosmology of weavers finds its most dramatic expression during privileged moments of social activity, specifically while making cloth for the shroud, it also informs the everyday practice of weaving. The everyday life of weaving refers as much to the execution of certain technical acts, as to the way these acts are organized during the day, the week and through different categories of people within the domestic household. For this reason, I will delineate the work of weaving into two areas: work in everyday life and work on the occasion of making cloth for the shroud (kafan).[1] The relation between the everyday and

[1] Cloth for the shroud is made in cases where death can be predicted, or when old members of the community insist on having their cloth woven. The occasion of making such cloth is an extraordinary event to the extent

extraordinary is found in the techniques of production. These techniques institute a way of doing and a way of representing in a mode that is both verbal and non-verbal. In contrast to the dominant theories of ritual which place them in the domain of the sacred, removed from the mundane life of man, I shall argue that the everyday and extraordinary enterprises of weaving are informed by the same cosmology.

Read together, both these domains of social life, the everyday and extraordinary, achieve a coherence of meaning in the day, the month and the ritual calendar year in a way that provides a discernible pattern of completeness to the practice of weaving. The attempted coherence of the everyday and the extraordinary is a statement both of the work of weaving and the community of weavers. Seen in this way, the techniques of weaving relate the story of cloth and the manufacture of this cloth on both everyday and ritual occasions.

Techniques establish rhythms of interactions that transform the physical movements of the body into externally visible and transmissible statuses and social relationships. In their widest sense, the techniques of weaving embody and personify the social world of both the weaver working at the loom and the worker providing services to it, and simultaneously express such a world. We may agree with Mauss when he says that technique makes available the concourse of the body and moral or intellectual symbols (1973: 70–88). In other words, the techniques of weaving do not mark an absolute separation between the mind and body of the weaver. It was common to hear weavers say that the clapping of the shuttle is the speaking of the word.

In the above sense, technique is not merely a means, but is, as Mauss says, an action that is effective and traditional. As an action that is both effective and traditional the technique of making cloth establishes relationships of equivalence between the body of the worker and the instruments of work. In effect, by having the property of 'standing for' or substitutability, a particular act and more generally, 'doing', becomes technique (Castoriadis 1981: 252–64). Second, relationships of equivalence include the possibility of repetition. A particular tool has the same use as another

---

that the category of work accentuates elements capable of transcending everyday weaving.

in that it can be replicated, or can have equivalents on the various occasions of its use. In this study such equivalence is not limited to material tools. The fabrication of particular social relationships in terms of codified attitudes, postures, gestures, practices and know-how is a technique by which the community of weavers frames its individual members. Obviously, weavers cannot be reduced to the techniques they employ in expressing themselves. Before the technique exists there must be a procedure by which it is posed as technique. But the artisan must have something which can be posed as technique, for, paraphrasing Castoriadis (1981: 267–8), the weaver cannot become a weaver without a loom. Nor can we say that the loom and other objects of work externalize an effective gesture, for this gesture becomes effective only by bringing the tool into existence: the capacity for extending and transforming the arena covered by technique is incorporated in its very organization. On the other hand, technique as stated above, cannot be realized without reference to signification, for then it would be a mechanical repetition of hands and feet. As weavers, the techniques they employ in making cloth reflect not only upon the quality of cloth and on the work that goes into its production, but also on the social relationships that weave cloth together.

THE SYNTAGMATIC AND PARADIGMATIC AXES IN THE ANALYSIS OF WEAVING

It follows that verbal representation and body gestures of cloth production are together co-present in the single act of weaving. These two, body gestures and verbal representations, form an ensemble of relations. The syntagmatic axis provides the sequential arrangement of weaving and details the body gestures of work; the paradigmatic axis establishes sets of equivalence by showing the basis on which one worker can be substituted by another. The social relationships in the act of weaving 'stand for' and 'serve for' something (Castoriadis 1981: 252). They are markers in that they relate linguistic activity, such as naming, or the uttering of sacred formulae to parts of the world, such as parts of the loom, or a particular sequence of acts in weaving. In this sense, the act of designation through linguistic utterances is specific. However, for such designation to be possible at all it must be capable of generality. This designation is called *zikr* by the weavers. It refers not only to the repetitive evocation of a sacred formula, but to naming in general, and to the memory of a work tradition embodied in naming. When these social relationships stand for they refer to classes of equivalence, of people, work styles, spaces and objects of work. All the occurrences of these classes have the same value at a given level. These occurrences are substitutable. This capacity for substitution is the basis of a paradigmatic relation.

To the extent that social relations of weaving and concomitant work styles, spaces and objects of work are substitutable, they refer to the technique of weaving. These social relationships, however, also serve for something. The signs designating objects and the work styles grouped around the objects are arranged in a particular combinatorial sequence, or a syntagmatic relation. Each sign is characterized by the possible combinations in which it may enter and by the potential use of related objects. The sequence that describes the syntagm begins with the category of a person at work on a particular object with specific instruments of work, including his or her body, located in a specified area of the hearth for the duration of the work.

Together, the two properties of standing for and serving for constitute the work of weaving. While the substitutability of instruments and people leads to the technique of production, the combination of people, acts, instruments and objects in a particular sequential pattern is given by the social structure of the community. Work, in other words, has three related components: the use of man-made objects, the observation of a socially defined ensemble of customs and the use of ordinary language. This is consistent with Heller's (1981) understanding of work. In this chapter I will show how this double characteristic of work (of standing for and serving for) is embodied by the weavers in making cloth. In the arrangement of sequences, in both everyday weaving and weaving for the shroud, the sanctity of social relationships is determined. I first discuss everyday weaving.

Weaving in Everyday Life

The diversity of everyday life is found in the way the time of weaving is structured. In everyday life, the act of weaving is characterized by its reversible temporality because the four stages of weaving are repetitive events in terms of the organization of skills. The four stages of weaving are repeated by the week. Coexisting with this reversible time is the career of individuals in the domestic group who occupy, at different points of their lives, specific positions. To this extent, the lives of individual members are reversible. Time here is the time of the body. Work in everyday life is ordered within the established periodicities instituted by the four stages of weaving and the various genealogical positions an individual occupies in the course of his life.

In everyday life, the temporal duration of the weaving process is understood through the weaving cycle which may be broken into two broad units: the sequence of the stages of weaving and the scheduling of each stage. The sequential pattern of weaving points both to the time taken to move from one stage to another and the order in which these stages occur. The weaving schedule locates each stage of the weaving cycle, and shows that each stage defines the personhood of a category of worker.

The four stages cannot occur simultaneously since each stage is processed according to an agreed upon career timetable of both the cloth and the person working on it. As a result each stage is temporally segregated. The basic contrast in the four stages is between weaving activities that occur in the night and those that occur during the day. Weaving alternates between day and night in that, first, sizing takes place during the day and is associated with light. Second, reeling begins during the day and ends after nightfall. The initial bobbin is always reeled in the darkest part of the hearth and is associated with darkness. Third, warping occurs in the afternoon and finishes late at night, but always in a way that it receives maximum light in the work shed. It is associated with light. Finally, work on the loom begins early in the morning and continues for a little more than four days. The loom is invariably placed in the darker part of the work shed. In its operation it embodies both the light and the dark. The weaving cycle is completed in a week.

Associated with the temporal segregation into the light and dark is the way each weaving stage creates a temporal boundary. Each stage cannot be subdivided or reversed. In assigning a person to work on each stage weavers recognize the irreversibility of each weaving cycle as well as associate a specific category of worker with each day of the week. The significance of this association is realized when cloth for the shroud is woven.

In everyday weaving the territorial area of the four stages is marked through the light and dark contrast. This contrast is apparently inverted when cloth for the shroud is made. Here, sizing occurs at night. Yarn moves from the mardana to the zanana. The warp beam is mounted at night and the weft members are made immediately after the morning call of the next day. This activity is timed so that weft members are made on a Friday, while the warp members are attached late on Thursday night (*Jumerat*), well after the last call. When cloth for the shroud is made the day begins on the evening of the previous day with the appearance of the moon. On this basis, the light and dark contrast for everyday weaving flows from the order established during ritual weaving. The light/dark contrast of everyday weaving must not be interpreted with the rising and setting sun, but should be seen in terms of when the day begins for the weavers during ritual weaving.

In making cloth for the shroud weavers assign both a specific place and time to each weaving stage. The yarn, cured at night, must be taken off the bamboo poles before the first call, irrespective of whether it is wet or dry. Reeling, too, is timed through the calls to prayer. Reeling must occur on the following day after sizing, before the appearance of the moon. All the concerned women must participate in it simultaneously. Like sizing, warping must be performed before the first and last call. Finally, wefting must occur immediately after the first call and be completed not later than the last call.

Based on this, the day is divided into two sections: activities occurring between the first and last call and those occurring within the last and first call. What is important is that the contrast between the light and dark is the basis of transcending the everyday work of weaving. This contrast is one that locates individual workers in an appropriate spatial and temporal environment and sacralizes this environment when cloth for the shroud is woven. The sacred environment of weaving is premised on repeating the community's deepest truths, embedded in the loom and other materials of work. Such work suspends the particularity of individual conditions and expresses obligations.

Work in everyday life reflects the divisions of the social structure and points also to the life cycle of individuals. At the outset, the basic division effected in the act of weaving is between males and females. Children, both male and female, below the age of six, do not engage in any weaving stage, nor is there a pedagogical transmission of the craft to them. At the age of seven, the male child is initiated into the community of weavers through a ceremony (*fatihah*) held in the month of *Bade Pir* (December–January in the period of this fieldwork). He begins work on sizing. With each passing year, until he is sixteen, his entry into the zanana of his father's hearth is limited. He comes of age as far as weaving is concerned when he stops sleeping in the zanana. He learns the work of the first three stages in his MB's dwelling. In contrast, after the age of six, the female child is not initiated into the community of weavers. Sometime in her sixth year she begins work on the first three stages. She is never formally taught how to weave. With every year, until she is fifteen, she provides her services in quilt making and is taught how to make them. By the time she reaches her teens her social interaction with males outside her hearth is curtailed. She, however, continues to enjoy unlimited access to her MB's and FZ's hearths until she is married. From the age of thirteen she wears the veil when she moves out of her village. Also, during this period she stops providing her services for warping.

Between the ages of seventeen and forty the male marries, raises children and establishes his own independent dwelling, with or without the ancestral loom. No Ansari adult publicly mentions his services for reeling. Outside his hearth, but within his kumba, the male in this age group provides his services for sizing and warping only under conditions of reciprocal obligation. Outside his kumba, but within his bihaderi, he sells these services under conditions of either reciprocal financial obligation or because of acute monetary pressures. In the last decade of this age-group he moves from warping to wefting.

A comparative age-group for women falls between the ages of sixteen and thirty-seven. This period extends from her marriage until her first child reaches the age of marriage. With few exceptions women work on sizing and reeling and a sizable number aid their father's hearth in warping. In this period a woman changes her hearth. She seldom sells her labour to non-kumba weavers, with the exception of her MB's hearth. After marriage and unless her husband is living in her father's hearth, she does not provide her services for weaving in her natal hearth.

In the next age-set for men (between forty and sixty) the weaver establishes his hearth, he becomes a grandfather and his sons fission off from his dwelling. As long as he weaves the ancestral loom is not transmitted. Once he stops everyday weaving he is considered to have retired. After retirement he transmits the craft of weaving to the second descending generation. In the first decade of this age-span he sells his labour to non-kumba hearths under dire financial need. Mainly, he provides his services as obligations within the kumba. During this period, too, he decides how much cloth is to be produced and from whom yarn is to be bought. He also decides whether a new loom is required. The addition of a new loom indicates his economic status both within and outside the hearth.

In the corresponding age-set (between thirty-seven and sixty) the woman head of the hearth provides her services for the first three weaving stages and, more importantly, makes quilts. For weaving, barring wefting, she controls the other three stages. I will now discuss each weaving stage.

Stage One: Sizing

In everyday life the relationship of the worker to sizing is seen from the perspective of the bride (dulhan) of the hearth and the children of the kumba. In hearths with more than one bride, the one married to the eldest man controls sizing. Yarn is initially cured just after sunrise in the compound outside the mardana, and is later shifted to a public area near a well. To cure yarn the bride must possess a complete mastery over the various counts of yarn. Because handspun yarn is not cured in the same way as machine spun yarn, she must know not only the different procedures for curing it, but also make sure that her hearth members are aware that she knows.[2] Initially, she checks the

---

[2] Handspun yarn is purchased from the Gandhi ashram, while machine spun yarn is bought from the city of Barabanki. In its raw form both types come in rolls, hollow in the centre. Machine spun yarn emits a pungent smell.

yarn for knots and breaks by running a part of the yarn between the thumb and forefinger of her right hand. Occasionally she whets her thumb to do so. She does not undertake a formal measurement but literally feels her way through it. After examining it she must voice an opinion which shows how well she has been tutored in her natal hearth. These opinions also express problems she faces in her conjugal hearth. If the yarn is handspun she loudly echoes the widespread complaint that it is of inferior tensile strength compared to mill yarn, and that handspinners have inflated its weight by concealing stones in the rolls. Complaints against machine spun yarn are not easy to voice, but often reflect on its brittle quality and dull lustre. The metaphors used in voicing complaints compare the yarn to faeces, particularly in terms of its smell, or to clay which must be given shape.

Having examined the yarn for its quality, the bride delegates work to the children of the kumba. The yarn is immersed either in cold rice water or in plain water and kept in an aluminium pot. After the yarn is soaked two children pound it alternately with their feet. Often the bride berates them for their laziness and lack of application. The comments made here by the bride refer to the body of the weaver at work on the loom. The children are urged to get on with the pounding since this is 'the learning of the feet (*pa'e ki ta'lim*)'. The treadles of the loom are referred to as the 'speed of the feet (*pa'e raftar*)'. The process of learning how to use the treadles begins with sizing. Among children she is spoken of, in her absence, as the one with the sharp tongue. The phrase I heard referred to the bride as 'tongue of the devil (*shaitan ki zaban*)'.

By midday the yarn is taken out from the aluminium pot and spread out by two children who hold it parallel to the ground. It is hung up to dry on two bamboo poles placed in the mardana of the yarn-curing hearth. The process requires delicacy of handling since the attempt is not only to preserve the maximum length of each strand of yarn, but also to ensure that the threads composing it do not come unstuck. The yarn is brought into the hearth before sunset and set up the next day. By noon of the second day the yarn is unwound from the bamboo poles and grouped according to its count. After it has dried the bride examines it for its colour — whether the different colours have run (*kachcha sut*), or remained in their former state (*pakka sut*).

Similarly, depending upon the quality of curing, she is spoken of as having a weak hand, or a strong hand.[3]

In the time taken for sizing the bride maintains a constant harangue directed at the children. This is in sharp contrast to her speech with other men and women members of the hearth and the kumba. She addresses the latter in the third person and speaks in monosyllables when spoken to, but may, in the next instant, sharply rebuke one of the errant workers. With her husband, especially, she maintains almost no visible contact, but in his presence, polices and commands the children more sternly. In supervising children she often conducts a conversation with the adult members of her conjugal hearth, but does not directly address them during this time. This conversation proceeds by highlighting an idiosyncratic bodily trait or gesture of a member of the kumba and then attributing it to a child worker. A phrase that came to my notice was, 'the torso is not a pole (*kamar khamba nahin*)'. It was directed at her husband's elder brother who, because of a spinal defect, was awkward in his gait. In making this statement the bride was expressing her anger at his attempts to constantly find fault with her ability to work in the house. On another occasion, with a split reference to this man's wife, she advised a female worker not to 'walk like a buffalo (*bhains ki chal*)'. It is as if the co-presence of the bride with adult hearth members must proceed by imprinting commands on children's bodies. Commands establish the dialogue with adult hearth members in that the tone used in orders expresses the bride's feelings about both the child worker and parents.

Here the bride is an outsider to her conjugal zanana, evidenced both in her area of operation, and in the kind of dialogue she establishes with other hearth members. By embodying this dialogue in the form of commands addressed to workers (devil's tongue, hawk eye, weak/strong hand) she telescopes a particular bodily relation to sizing and the weaving process in general. This relationship is one of temporal and spatial exclusion in that the bride works outside the male part of the house. Until her first child she cannot work in the zanana. This relationship of exclusion

---

[3] The weak colour of yarn is often compared to the Khan Saheb woman in parda. Her complexion is *kachcha* because, as the weaver reasons, she is either sequestered in her zanana, or is hidden from the sun's rays due to the burqa she wears in public.

is marked by the bride's behaviour with working children, on the one hand, and her practice of parda in the presence of adult members of her hearth and the kumba, on the other.

Stage Two: Reeling

In this stage the woman head of the hearth and her daughter (beti) or son's wife (bahu) handle the yarn and the implements of work. Here yarn is reeled into bobbins. The first bobbin, called *kunda*, is reeled by the woman head in the part of the zanana where the nuptial bed (*khatiya*) is placed. The kunda is always reeled while in a squatting posture. In former times this posture was the preferred one for delivering babies. In reeling subsequent bobbins the woman sits on the ground with one leg tucked under the other which is outstretched. I have not seen any weaver sit this way. Bobbins are reeled after the yarn is taken off the bamboo poles. This occurs by about noon of the second day of the weaving process. Often the woman head reels the second and subsequent bobbins in the quilt room and occasionally in the compound of the mardana.

Initially the dried hanks of yarn are loosened from the bamboo poles and wrapped around a simple spinning-wheel. The wrapping is done by the woman head. The wheel is a post with four arms pivoted at the top, not unlike a turnstile. It consists of a vertical wooden axle into which four pairs of rods, crossing each other diagonally, are horizontally inserted and tied in place. Two cords run in a zigzag formation between the eight corners. The left hand of the woman operates this turnstile, which rotates in a naturally-holed stone. A hank of cotton is wrapped around each of the four arms of the axle. The ends of the threads are gathered together and fastened to a fifth, somewhat larger turnstile, which is now made to revolve rapidly so that yarn from the smaller turnstile is wound around it. After the winding process an upright spindle is set up between the larger and smaller axles. This spindle is called *indi*, also a colloquial term for the penis. The indi is supported by a slanting rod with a barb, against which the axle of the turnstile leans. A bobbin is mounted on the spindle axle and while the right hand turns the crank, the left places the yarn threads on the spindle. Subsequently, while the yarn is being rewound it is twisted between the thumb and forefinger. In this way yarn acquires tensile strength. When the bobbin is fully reeled, the four strands of yarn, termed *bal,* are snapped off, and the full bobbin is replaced.

As in sizing, the woman head must know the complete details of the process of mounting the yarn and operating the wheel. Bodily idioms provide a mode of access to this knowledge. The first bobbin is reeled by the woman head in complete privacy because it is believed that the birth of cloth occurs from the point the kunda was reeled. Yarn of off-white colour is always reeled the first time. Prior to its reeling the woman head sets up the implements by herself in the area of the nuptial bed. While reeling she must not only sit in a squatting position, but also ensure that the threads snap as infrequently as possible. Yarn threads snap most often when they are being placed on the spindle (indi). Once reeled the kunda is examined for flaws. Accordingly the quality of cloth that will be woven is determined. Male weavers compare the setting up of the apparatus near the nuptial bed to the ritual of *kanch,* which occurs just before a marriage is consummated. Kanch refers to a period when the groom's eBW (elder brother's wife) or his FZ escorts the bride to the nuptial bed and formally introduces the groom to the bride. In this context, kanch means hymen. As far as weaving is concerned, the phrase I heard was, 'the yarn is in kanch (*sut kanch men hai*)'. The fully reeled kunda is examined by commenting on the strength of the yarn. The woman head is often questioned about the number of times the thread snapped while the kunda was being prepared. A typical reply is, 'I lost hold of it twice (*Humse due bar chut gaya*)'. Accordingly, the strength (bal) of the yarn is determined. Yarn strength is determined in terms of its ability to remain unbroken. This strength is formulated through a sexual metaphor in that the snapping of threads is equivalent to the rupturing of the hymen.

After the first bobbin is reeled the apparatus of reeling shifts from the area of the nuptial bed to the quilt room, or the compound of the zanana. During this time the woman does not renegotiate her status as much as restate it by instituting the code of parda. This is apparent from the way work is done in the zanana. Here, work is done by more than one woman, who alternate on the implements of reeling. The woman head exercises minimum supervision and is often engaged in household chores.

This is the time for gossip and kumba politics and an occasion for matchmaking. I have heard male weavers say that the conversation of women in the zanana is louder than the clapping of their shuttles. One such phrase is, 'In the time of the bobbin the shuttle does not sit in the ear (*Bobne ke waqt men nal kan men nahin baithat*)'. Weavers, however, do not have direct access to the conversations and tensions that prevail in the zanana at the time of reeling. The stock reason given is, 'while speaking we cannot burst our vocal chords (*hum gala phad ke nahin bol sakat*)'. These sarcastic remarks refer not so much to the conversation of the women of the zanana as to the fact that all adult men are, in the duration of reeling, barred from entry into the zanana. Parda does not so much seclude women as exclude men.

Unlike sizing, parda is not observed by a single individual in relation to all adult hearth and kumba members. Parda in reeling is centred around a group of women at work in a specific area of the hearth. Among themselves women workers do not maintain prohibition of sight and speech, but collectively and deliberately, ignore any man who strays into the zanana. Through parda what is being stated is that the reproduction of people over time in a weaving hearth is an authoritative resource which depends upon the specific arrangement of workers at work with certain materials in specific places.

Stage Three: Warping

Bobbins are attached on to the warp beam while reeling is still in progress. After a sufficient number of bobbins are reeled (about eleven to fourteen) they are taken to the work shed (karkhana). This houses both the warp beam and the loom. Beginning in the late afternoon of the second day warping carries on well into the night or early morning of the third day.

Just as the zanana is the domain of the female head of the hearth, so is the work shed of the male head. Usually he does not mount the bobbins himself, but supervises this task, which is carried out by his son/s or younger brother/s with or without contracted labour. The warp beam, placed beside the door of the work shed, is made of two vertical wooden posts, about eight feet long and dug into the ground. The two posts are connected at the top by a third which is placed horizontally between them.

Below the horizontal post thin strands of wire, running horizontally, connect the two posts. These wires are arranged from the top to the bottom of the posts. Usually between six to eight wires are employed by the weavers. The bobbins are arranged according to the colour combination and count of the yarn on these wires. While fixing bobbins on them the worker remains in a standing position to ensure they are firmly attached. He must have a carefully measured mental picture of the design and colour combination to be woven, as well as the quantity of yarn employed in its making. A sense of balance and proportion are important characteristics of the mental picture. After the bobbins are fixed on the warp beam their yarn threads are drawn out and attached to the loom. Depending on the count and weight of yarn as well as the required design, such threads are attached to the loom at varying distances. This is perhaps the most tense moment of the weekly cycle for yarn threads often snap while they are being attached to the loom. More important, the precision of the design and colour combination is determined at this point and if the bobbins are not properly located on the warp beam the entire process begins anew.

The male head of the hearth exercises strict supervision over the workers. While he speaks directly to each individual in the form of commands, they refer to him either in the third person or by a genealogical term. Unlike during sizing, commands refer to the immediacy of the task at hand. Often the male head demonstrates how the work is to be done, taking care to show the fluidity of his upper body movements and economy of effort. These actions show both his skill as a weaver and communicate how well he has been schooled by his father. A particularly skilful head weaver, whose prowess in warping is well known, is spoken of as 'strong of speech (*kalam ka pakka*)', and is often called 'wise weaver (*bunai ka hakim*)'. Speech is embodied through work. In its widest sense this embodiment refers to the proportionality and combination of letters (*kalimah*). This will be explored later.

In addition to the attachment of bobbins, skill in warping is evaluated by the proportion, colour combination and design of both yarn arrangement and cloth. In everyday life this measure leads to the computation of quantity. We will see, however, that in weaving cloth for the shroud the numbers have a significance that is qualitative and not reducible to numbers.[4] When the yarn threads are drawn from bobbins and attached to the loom, the workers, under the guidance of the head weaver, do not use a measuring instrument to gauge the length of the threads being drawn out. Experienced workers judge the distance by sight, while novices determine it by using the length from the tip of their outstretched finger till their elbow joint as measure.

The colour combination of the design to be represented on cloth is based on an equilibrated balance between the light and dark. The warp beam is placed near the door of the work shed so that it receives the maximum light of the sun. After sunset a lantern is placed near the beam. One reason for situating the lantern thus is pragmatic. In attaching the threads to the loom the warp members often snap, either because they have been reeled too tautly, or because threads of different bobbins have intertwined and need to be separated. A second reason is that the light of the lantern releases the energy of warp members imminent in the bal, thus achieving a correspondence between the spiritual and corporeal worlds. The visible aspect of the strength of warp members — their tensile strength and ability to stretch without snapping — presupposes an equilibrium with an invisible counterpart, given in the two numbers seven and twelve. This energy must not only be combined in the right proportion, but the combination must be made visible to our 'eyes of the flesh'. At the two extremes of this combination we have light and darkness, and between them a series of proportions which manifest the two extremes: the combination between light and colour, between the strength of warp threads and their ability to stretch, between the size of the design and that of cloth. This makes visible the balance between light and dark. Furthermore, this visibility is repeated by the week, for a true combination between the light and dark manifests itself independently of the individual weaver. This repetition involves the idea of a return to equilibrium by means of the movement of time as cyclical.

By the cyclical nature of time the younger generation takes the

[4] The smallest unit of measure is called 'hand' (*hatthwa*) and corresponds to a yard (*gaz*). Depending upon the weight and count of yarn, between eight to six hands make one-fourth (*chauka*) of a roll of woven cloth (*than*). The average weekly productivity per weaver varies from between 65 to 70 yards.

place of the older. This replacement finds its genesis in the pedagogical transmission of the craft from father to son. Weavers refer to this relationship as similar to the teacher/pupil (*pir/murid*) one. For the purposes of learning warping and wefting the novice among his colleagues refers to his father as *pir sab*. Here the father, through words and non-verbal symbols, schools his son into the craft of weaving by showing him how to mount bobbins on the beam and make warp members. Verbal symbols are mainly propositional, while gestures mime the message the weaver wants to make manifest. Propositional statements of the head weaver effect a relationship of resemblance between the task at hand and the novice's attempt to carry out the work. Non-verbal symbols, on the other hand, replace resemblance with identity insofar as the head weaver demonstrates mimetically the work to be done. In both instances the novice learns warping by the use of his body. The mimetic display of learning is considered the work of a poor craftsman, for weavers believe that quilts, not cloth, are made thus. Involved in the tutoring are various injunctions: the murid must obey his pir, but not by imitating him; in the presence of the pir he must never touch the loom; he should neither eat nor drink in front of the pir, unless commanded by him; finally, he should not attend to any work other than asked of him by his pir.

Stage Four: Wefting

Wefting begins immediately after the warp members are attached to the loom. The male head of the hearth starts working in the early hours of the third day after his bath. He must not eat anything before he touches the loom. As soon as he sits down to work on the loom he recites the first three or four du'a (supplicatory prayers)[5] associated with the relevant pieces of the loom. In his absence his younger brother or one of his sons will work on the loom and observe all the injunctions associated with such work except one: he will not recite the du'a. Beginning on the third day, wefting carries on until midday of the last day of the week.

To make cloth the shuttle of the loom is thrown through a

---

[5] Supplicatory prayers are found in the *Mufidul Mu'minin*. The text will be discussed in the third chapter.

shed, which is a horizontal narrow passageway traversed by the shuttle from the left to the right and back again. The weft thread just thrown is now drawn in at its centre point toward the comb. As soon as the shuttle passes clear of the shed the comb is brought forward, the weaver grasping it by its upper baten with his left hand. By this the inserted weft is forced into its proper place. The weaver now changes the shed and throws the shuttle back with the other hand. The use of the right hand in throwing back the shuttle arises only if the loom has a single shed. In such instances, the weaver uses his right hand to work on both the grip and the shuttle in its movement from the left to the right of the loom. The warp is used in a dry state. To ensure there is no dampness weavers subject the warp members to the heat of a lantern which is placed close to them.

In teaching his son how to weft, the male head of the hearth explains the spatial division of the loom and the corresponding placing of hands and feet. The head weaver recites the first three or four du'a, but never asks his son to repeat them. The combination of instruction and recitation establishes a relation between the rhythmic pacing of hands and feet and the phonic pacing of prayers. These prayers are said to be the 'inner eye' of weaving and all verbal instructions directed at the learner are emblematic of this inner eye. The correspondence between the movements of the body and the recitation of prayers is said to rest upon the precision of the colour combination and design of woven cloth. That is, the synchronization of hands, feet and speech is determined by the equilibrium of colours and designs. Thus a master weaver is spoken of as someone who has the word and is strong of speech.

This equilibrium rests on the balance of four. The number lends itself to various combinations, but is rooted in the balance between the light and dark, which in turn, leads to four colours: white, red, yellow and black; four types of designs given morphologically by the straight upright line, the circle, the triangle and a horizontal serrated edge (which on cloth is manifested by arabesques). Finally, it leads to a balance in nature given in the four elements, earth, water, fire and air, and their combination in the four stages of weaving. The full implications of this balance will be explored in wefting for the shroud. Here I will explore the range of permitted colours and designs.

White, red, yellow and green are the colours used in making cloth in everyday life. Green, the weavers say, is not of the same order as the four elements mentioned above, since it refers to the world of the soul and it is impossible to find its actual existence in the corporeal world because of its intense subtlety. In the everyday world of weaving the green of the Prophet's garden is manifested by the colour black. Weavers will never combine black with any other colour while making their cloth. The usual colour combination found on cloth is a white base with green or red arabesques running through it. A second type of design is a white base with yellow or red squares, rectangles or rhombuses, located in the centre or on the borders. A third common design is a white cloth with leaves or flowers (usually of a local mango tree), and trellis woven in yellow or green on the corners of it. Finally, as is most common, we find cloth that is plain white and has no designs on it.[6] Under no circumstances will the weaver represent living creatures.

What is the relation between this colour combination and the four designs with the body of the weaver and worker? Schematically, the straight upright line corresponds to cloth that is of white colour; the circle to the combination of white and yellow, or white and green; the triangle to white and red, or white and yellow; the serrated edge to white and green, or white and red. In everyday life the straight upright line corresponds to a uniform and transparent wefting, in that each point on the cloth is visually the same. For that reason it appears surface deep. The circle corresponds to the wefting of leaves and flowers in that the visual appearance gives the appearance of roundedness. The triangle harmonizes with sharply angled geometric figures such as the square and the rectangle, while the serrated form agrees with arabesque in terms of angles and dents.

Corresponding body metaphors describe the designs with their relevant colour combinations. The straight line does not have body metaphors associated with it. In its reproduction by the artisan the idea is not so much to evoke a corporeal presence as it is to the colour's sacramental value. The signification of white

---

[6] On occasion, when the weaver makes cloth for the market, such as saris or lungis, the colour combination he employs remains more or less intact, though the design may be more elaborate.

represents all possible combinations. The body metaphor evoked by the circle is phlegm (*balgham*), characterized by its congealing inertness and lack of cohesion. It has the modality of the colour yellow. The opposite of such a metaphor is provided by the design of the triangle, which evokes the characteristics of semen (*dhat*), described by its swelling, expanding and ascending motion. Finally, the design of the serrated edge evokes the teeth, characterized by the compact cohesiveness of the parts. The modality is of the colour green.

The balance of four structures the time of everyday life in weaving. This equilibrium works simultaneously in two ways: it structures the day as well as the ritual calendar into three equal parts. Simultaneously, it associates categories of people with such structuring. Each quarter of the day and of the ritual year is referred to as *chauka*. The first quarter of the day, i.e. from the time the weaver sits down on the loom till he has his breakfast (*nashta*), is characterized in terms of the body of the weaver as a period of renewal in which the balance of day and night is broken in favour of day. The body is thought to be at its strongest and most fertile. The category of the bride (dulhan) is associated with this period. According to a belief widely held by weavers, she is most amenable to impregnation during this period. The first four months of the ritual calendar correspond to this duration, i.e., *Id*, *Khaliq*, *Bakr Id* and *Muharram*. Marriages occur most often in these months.

The interval from breakfast till midday corresponds to a period of abstinence, fatigue and lethargy, for in the noon hour the body becomes stifled and parched. This division marks the start of the day's complete domination over night. The category of the female head is associated with this segment. She is thought of as frugal in her diet and sharp with her tongue. The next four months of the ritual calendar harmonize with this time, i.e., Chahullam, *Barawafat*, Bade Pir and *Madar*.

The third division of the day, the interval extending from the midday meal till the last call, is a time when the day's production is gauged and when the order established at the beginning of the day comes to fruition. This is the time of the menfolk of the hearth, of the azan, and of restful sleep, marked by the emerging dominance of the night. The metaphor is of the body in repose. The last four months of the ritual calendar, i.e., *Maqdoom*, *Rajab*,

Sabrat and *Roza* correspond to this interval. It is not coincidental that on the ritual of Sabrat old people of the community who think they have led a complete life stay awake at night praying for deliverance.

There is a fourth period of the day characterized by the sadness and desolation of living bodies in the sensual world. According to one weaver's view, the prevalence of night over day initiates a period of uncertainty ending in catastrophe, a time when the devil manifests himself. For this reason weavers speak of this period as *qayamat* (resurrection and judgment), when the account of those who are to die shortly will be written. In everyday life this moment of judgment is negotiated during daylight by assiduously carrying out the work that describes one's status in the hearth. This is also the time when ancestors commune among themselves and lament their corporeal life in this world. This is finally, the time of the exile and of the stranger (*ajnabi*). The category of the stranger lends itself to various connotations. In this context, the appearance of the stranger, when the rest of the world is asleep, is the condition of the transformation of event into symbol. This event is the person enacting the modality of the four stages of weaving.

WEAVING FOR THE SHROUD

Before I describe the four stages of weaving it may be useful to briefly discuss the death ritual. My discussion refers to instances where death can be predicted or when an old member of the community has his cloth for the shroud woven before he or she dies. In the latter case such cloth is made during the month of Sabrat after the concerned person has stayed awake for one night praying for deliverance. Cloth for the shroud is not made when death is accidental. For the Ansaris making cloth for the shroud takes precedence over the rituals performed during the burial of the corpse because through such production they re-establish contact with the sacred tradition of weaving. It was put to me thus: 'In making cloth for the shroud the Ansari becomes nurbaf'.

Haji Ghulam Nabi's wife, about 67 years old, died on Thursday, 19 December 1985, of an unspecified stomach ailment. She had been removed to the hospital situated in the village ten

94 • *Work, Ritual, Biography*

days earlier, but died in her house. She was ill for a long while and the *Haji*'s dwelling had reconciled itself to her death. Ghulam Nabi had two elder brothers, the older of whom had died five years ago (Ghulam Nabi himself died in 1986). The Haji's surviving brother, Mushtaq Ansari, is the eldest surviving member of this particular agnatic line. The Haji had two daughters, both of whom lived within the kumba. His youngest daughter is married to her classificatory MB's son (FBWBS or father's brother's wife's brother's son), who is a *ghar jamai*. The eldest daughter is married to Mushtaq's youngest son. The diagrammatic representation is shown in figure 1. (See categories of persons involved in making cloth for the shroud).

FIGURE 1: CATEGORIES OF PERSONS INVOLVED IN MAKING CLOTH FOR THE SHROUD

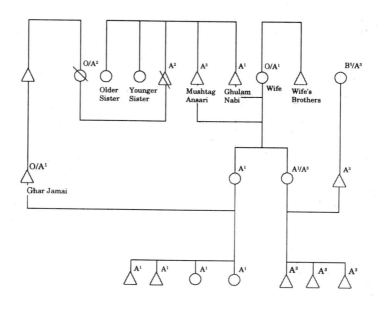

With the hospitalization of the Haji's wife, A^1 was left without an effective female head. Until the weaving of cloth was complete, Mushtaq's wife would act as the surrogate head of A^1. Throughout

the duration of making cloth the kitchen of A^1 was not lit, although Ghulam Nabi's wife had not as yet died. The Haji was, however, told by the hospital doctor that the patient did not have much longer to live. His wife was brought home ten days after she was admitted to the hospital. The two male elders of the agnatic line decided that cloth for the shroud would be woven on the coming Friday. Meanwhile Mushtaq's youngest son, married to the Haji's daughter, carried word to his father's sisters, both of whom lived in separate kumbas. Ghulam Nabi's WB, living in another village, was informed by the ghar jamai.

The patient's brother arrived on Wednesday night. After washing his hands and feet and offering prayers in a nearby masjid, he began to wash the yarn in rice water. Each separate strand of yarn threads was bathed, not merely allowed to soak in rice water. The pot which contained rice water was thoroughly rinsed before this liquid was poured into it. Two hours later the yarn was taken out for drying. The yarn was bathed outside the hearth but dried in the zanana, where it was hung up on two bamboo poles, with lanterns placed very close to the drying yarn.

By noon of the next day (Thursday), the yarn still damp, was taken out for reeling. Early on Thursday morning, Ghulam Nabi's two sisters arrived with cooked food and immediately went into a room in the zanana where his wife was lying. This is a moment filled with anguish because the person for whom cloth for the shroud is being made is certain about the nature of the visit. In this instance the Haji's wife had apparently been pleading for some time that her 'cloth' be made.

To reel bobbins the women of the haji's generation (his two sisters and Mushtaq's wife) participated in the work. Approximately eighty-five bobbins were to be reeled. The three executed this task simultaneously. Three ordinary spinning wheels were operated, two of which belonged to Ghulam Nabi's hearth (one of them was the property of the ghar jamai) and the third came from Mushtaq's hearth. As opposed to a single individual working on the first, third and fourth stages, reeling is done collectively.

Warping was done by a worker of a generation lower than the Haji's. This person is always a member of the same kumba as that of the hearth for which cloth for the shroud is made. Weavers say that this person is the FBS (bhatija) of the concerned hearth. When I asked who made bobbins for the shroud the inevitable answer was, 'so and so's bhatija'. So and so refers to the male head of the hearth for which this cloth is fashioned. In response to the next query, 'If there is no bhatija available in the kumba?', I did not find the answer I thought I would ('He is related as a bhatija would be related to him') but the more emphatic 'this can never happen in the biraderi'. In the present instance, Mushtaq Ansari's son attached bobbins on to the warp beam. The beam had seven wires placed on it, on which twelve bobbins on each wire were placed. Lighted lanterns were situated close to the beam and would not be extinguished until after the cloth was made. The Haji's BS worked in complete privacy. Early on Friday morning, immediately after the first call, Mushtaq Ansari, after bathing, sat down at the loom to fashion the weft members. The cloth must be made before the last call and at one sitting. While weaving this cloth Mushtaq drank only water, and would intermittently chant the prayers mentioned in the *Mufidul Mu'minin*. A branch of a local mango tree lay to the left of the loom near the grip. Once the cloth was woven it was not rolled, but left to lie where it had fallen. After Mushtaq had finished weaving the entire cloth he gathered it and took it into the zanana. It measured approximately 6 by 8 feet. The loom on which such cloth is woven is the one that is transmitted.

On such occasions work is obligatory in that there is an institutional compulsion to meet the requirements for making cloth for the shroud. This means that, first, none of the members in whose dwelling such cloth is being woven will participate in this work, except in cases where the head of the dwelling is also the head of the patrilineage. Second, the allocation of members in each of the four stages points to the connection between the household, kumba and bihaderi. Yarn is cured by an affinal relative of the male head of the hearth, or his MB. He is always a member of the same bihaderi living outside the kumba. Reeling shifts the focus to the women of the kumba and the bihaderi. Warping maintains the generational difference within the kumba while wefting affirms the bond between the hearth and the agnatic line. Third, weavers assume that every individual member of the community will, at some point in his or her life, be called upon to provide his or her services for this work. Obligation should not be construed as long term dyadic reciprocity, but as the genealogical place an individual occupies in his agnatic line. As a corollary, this work expresses a continuity with the hearth's ancestors and with the generation to come, seen when the loom is transmitted. As shown above the social structure of the community inheres in the production of cloth for the shroud. The significance of this production is not limited to the social structure of the community, but extends beyond it to express a system of cosmic worship by privileging the body techniques of the workers and the body of the corpse. This is examined through the four stages of weaving.

Stage One: Sizing

When cloth is produced for the shroud the type of personnel who work on the four stages are different from those who work for everyday production, though there are obvious relationships. In sizing the MB of the male head of the hearth where this cloth is being made carries out the task. As in everyday production, yarn is bathed in rice water outside the hearth, but dried in the zanana. Further, in the former instance sizing begins immediately after sunrise. For the shroud yarn is cured after sunset. In both cases the person who cures yarn is related by marriage to the specific hearth. Both are outsiders in a proximate relation with the concerned hearth. Finally, while in everyday production yarn of different colours and counts is cured, here white yarn of a single count is cured.

Sizing here is better understood by placing it within the light/dark contrast. Within this contrast bodily idioms provide a normative discourse, both of the work process and the type of people involved in it. While talking of shroud production a head weaver told me that uncured yarn is naked, waiting to be clothed with the sun's rays. He added that the first contact with this yarn, in the form of cloth, makes the body of the deceased free from matter (*tajjarud*). To pursue this point I will show the importance of the two areas of the hearth where sizing is done.

The yarn is cured in the compound near the mardana and inside the zanana. When yarn is hung up on two bamboo poles the zanana must be brightly lit, with lanterns placed close to the wet yarn. The idea is not only that the heat from the lantern will dry the yarn, but also that it will be clothed in light. The phrase used is, *latifah ki roshni*. In this context the phrase refers to the light of the soul illumined by Allah. Such yarn, in turn, becomes a source of light which illumines and makes visible other objects by virtue of its intrinsic nature. In answer to the question — why is yarn bathed in an area without illumination? — I was told by the head weaver that the colour of darkness has to be made visible. By darkness he meant not only the shadows thrown up by light, but also the colour of the corpse (*murdah*). The quality of colour of the corpse is manifested only by a more powerful light (*nur*). In the same way, the light of the yarn in the mardana is made visible by the light (*roshni*) of the lanterns in the zanana.

It is important to distinguish between colour (*rang*) and light (nur or roshni). Darkness for the weavers is not the absence of light. It manifests itself as such and this manifestation is integral to the dead body's nature. For this reason the body of the departed must be clothed with nur so that it manifests its intrinsic colour. Thus, light is the cause of the manifestation of colour and colour is, as far as the body is concerned, light made corporeal. Also, the notion of the body is not limited to a physical body of this world since its colour is illumined by the light of the soul. In effect, we have spirit and body, light and colour, distinct yet inseparable, the one manifest by the other.

Sizing for the shroud also involves the living community of weavers. More specifically, it points to the signature of the MB, who moulds the elements used in sizing. I use 'signature' as much to emphasize the uniqueness of each stage, as to show that each stage is authored by a set of rules that together constitute the body of the deceased. The modality or style of the signature is given in the conjunction between water and earth together related to fire and air. While bathing yarn in rice water in an area without illumination, the MB represents a dormant femininity (inert earth) and humidity (rice water). When the yarn moves into the zanana the MB represents dry heat (a combination of fire and air) which, I was told, corresponds to an awakened femininity. This alchemical point was explained to me in the following way: in any serious quarrel between husband and wife two arbiters are chosen who belong to the husband's father's household, and the wife's father's household respectively. The arbiter chosen from the husband's side is hot and dry (fire and heat), while that chosen from the side of the wife is cold and wet (earth and water). The reconciliation between the two opposites leads to the stability and nuptial union (*'urs*) between fire and earth. For the union to be perfect the two partners who constitute it become four. In this way, yarn must be cured, first by water placed on earth and then by fire while it is suspended in the air. The two pairs, earth and water and fire and air are assigned separate areas of the hearth. The mark that distinguishes the first two (i.e., mardana and zanana) from the last two is the one the dulhan negotiates to become an insider to her conjugal hearth.

Stage Two: Reeling

Reeling for the shroud begins, as it does in everyday production, in sunlight. There are three differences between the two. First, yarn is reeled when still damp. Completely dry yarn is never reeled. Second the entire yarn must be fully reeled before it is attached on to the warp beam. Third, the personnel who do this work are not members of the hearth in which death is imminent, though they are all women of the kumba of which this hearth is a member and the bihaderi. In addition to the sisters of the male head of the hearth, these women belong to the same generation as the person who is the agnatic head. The eldest sister of the male head of the hearth reels the first bobbin, but not in the area of the nuptial bed. As in everyday production, bobbins are reeled in the quilt room.

In producing cloth for the shroud, white yarn of a single count is cured and reeled and is thought of as being clothed with the sun's rays. Yarn so clothed acquires a material significance that it lacked in sizing. Further, while in sizing a union is suggested between dry heat (fire and air) and cold humidity (water and earth), in reeling the suggested union occurs between heat and humidity, on the one hand, and cold and dryness, on the other. This is the modality of the signature in reeling. The corpse is embodied with the imprinted signature of the women who reel, especially the sister of the male head of the hearth. Simultaneously, the corpse informs the other half of the signature. The imprinted signature becomes visible only when it is tinted by earth, for the invisibility of the radiance of cured, but not reeled, yarn is due to the extreme intensity of its whiteness, not due to its obscuration. For this reason, yarn must be damp when it is reeled.

In turn, this tincture makes the true nature of the corpse visible (*murdah ki roshni*) to the living.

Heat and humidity are found both in the yarn being reeled and in the personnel who reel them. The yarn, though subjected to the light of lanterns, is damp while being reeled. This combination results in the swelling of the corpse when it is clothed in reeled yarn. The analogy drawn is of water put to boil: steam always ascends while water is being boiled. The issue is of making this steam visible. Steam manifests itself in the body of males in the form of semen which combines heat and dampness. In women steam is found when they lactate. The swelling of the corpse results in its re-acquisition of carnality (*nafs*) that it had lost at the moment of death. While heat in the sense of fire postulates the colour red, the expansiveness of the substance assumes the colour white. The result is a mediating colour, which is yellow (the colour of semen and milk). Cold and dryness constitute the corpse and together characterize the body of the deceased. The colour of the body is thought to be black, the colour of terrestrial earth. It is not only that blood ceases to flow and the body becomes cold but that it cannot engage with the visible and carnal world. While dryness postulates the cohesion of the parts of the body, the coldness of the corpse inaugurates a movement into a subterranean region of the earth.

When heat and humidity (white) conjoin with coldness and dryness (black) the corpse is restored its carnality, but not in the visible, terrestrial world. Though made of flesh and blood it is invisible. The conjunction between white and black is found in the union between the implements of reeling and the yarn that is reeled, between the indi of the spinning wheel and the kanch of the hymen. The question remains: in production for the shroud why is the first bobbin never reeled by the side of the nuptial bed? This question may be posed in the context of the relation that the personnel of reeling share with the dying person.

In the weaving process this relationship depends upon the maintenance of blood and affinal ties, given to each kumba through the term zikr. Zikr refers to the naming of everything in this world, including social relationships. This naming occurs through a repetitive recollection of a specified formula, which evokes the style of the signature, or more appropriately, a knowledge of what such style calls to mind. The structuring of memory and the organization of reeling are embodied in this signature. In the case of reeling, the quality of zikr that the reelers share with the dying person replicates the nurturing relationship of a mother to her child. In providing the child with her milk the mother ensures the continuity of the child's life after she herself has died. In the same way the corpse must be made ready for life in the non-visible world after it has died, signified by reeling yarn that is off-white in colour. For this reason women do not reel by the side of the nuptial bed. The significance of zikr lies in that in the world of both the living and the dead there must be a name that corresponds to the state of that world. In this sense colours encode the memory of a certain substance which can only be recalled by the modality of the signature. The name of each such signature (for eg., *bahen*, *bobne* and zanana in the case of reeling) recollects obligations that certain types of people must perform.

In sizing and reeling the modality of the two signatures is not used in the same way. In sizing, yarn has the quality of both hot dryness and humid coldness; in reeling, reeled bobbins have the quality of humid heat, and the corpse is infused with the quality of cold dryness. Weavers say that through the modality of cured yarn the corpse becomes visible as cold (water) dryness (air). For this reason, reeled bobbins manifest the hidden nature of the corpse by contrasting its cold dryness with humid heat.

Stage Three: Warping

The bhatija of the male head of the hearth attaches bobbins onto the warp beam and makes the warp members of cloth. The bhatija must come from outside the hearth. Bobbins are attached in much the same way as they are in everyday production. The significant differences are that, first, bobbins of different colours and counts are not attached since cloth for the shroud is plain white and of a single count. Second, a qualitative importance is placed on two numbers, seven and twelve. The number of bobbins that constitute one horizontal row of the warp beam will always be twelve, while the number of wires that run from the top to the bottom of the beam will be seven. In making cloth for the shroud the bhatija works with seven rows of bobbins and each row has twelve bobbins. This leads to a balance between the temporality of spiritual time and of everyday time. Third, the bhatija fixes the bobbins on the beam after the last prayer call, when the sun has set. Fourth, before entering the work shed, the bhatija must wash his entire body (*ghusl*) and rinse his hands in the same rice water that was used for sizing.

The category of the bhatija supplies the modality of the signature of warping. While in reeling earth and air are combined with water and fire, here a conjunction occurs between air and fire, on the one hand, and earth and water, on the other. Warping is distinguished from sizing in that while in the former the union is between the two pairs, fire and air and water and earth, in that order, in warping such a union occurs by reversing the order of each pair. What we have is not hot dryness and humid coldness as suggested by sizing, but dry heat and cold humidity.

After the bobbins are fixed on the beam they are subjected to the light and heat of a lantern. This results in the cohesion of warp members and a unidirectional movement towards one centre. On the beam the bobbins are suspended from earth. They are dry. When they come into contact with a lantern they have the quality of heat and light. An example of such a configuration is the pen (*qalam*) used in writing the word of God and also the account of the departed. At the same time warp members are infused with the signature of the bhatija and are in contact with an inert loom placed on earth. Such contact makes warp members cold and humid and results in the absence of upward movement. Warp members are cold because they are in contact with a lifeless mass (loom), and humid because the bhatija's hands, rinsed in rice water, attach bobbins on the beam and fashion the warp members. Here, warp members have the quality of darkness, exemplified by phlegm (balgham). The yarn, in sizing and reeling, worked upon by women in the zanana, must receive the imprint of the men at work in their part of the hearth. This imprinting reverses the modality of the signature of sizing and reeling and points to a male belief that the relationship between sexes is symmetrical and reciprocal. In the specific case of warping the zikr of the modality of the signature shows how temporal succession, in the sense of the movement from sizing to warping, is stabilized in the order of spatial simultaneity given in the warp beam. This is seen in the significance of the two numbers, seven and twelve. That is, the first three stages are materially represented on the beam before the warp members are made.

The number seven locates certain attributes and embeds them in categories of persons in a hierarchical order without applying them to any single individual. The order shows the representational significance of weaving organized along the materials of work. The seven attributes are: life (*hayat* or *zindagi*); intelligence (*aql*) and sometimes knowledge (*'ilm*); power (*qudrat*); will (*iradah*); speech (*kalam*); letters (kalimah); pen (qalam). Corresponding to these seven attributes are the seven days of the week, which manifest each attribute and seven categories of persons associated with the week. This is shown in the following chart.

THE SIGNIFICANCE OF THE NUMBER SEVEN

Attributes	Day	Category of Person	Material
Life (hayat or zindagi)	Monday (Dushamba)	Bride (Dulhan)	Yarn
Intelligence (aql) knowledge ('ilm)	Tuesday (Mangal)	Mother's Brother (Mama)	Yarn
Power (Qudrat)	Wednesday (Budh)	Father's Sister (Phupphi)	Bobbin
Will (iradah)	Thursday (Jumerat)	Son/Brother's son (Beta/Bhatija)	Bobbin and Warp Beam
Speech (kalam)	Friday (Juma)	Father/Father's Brother (Walid/Chacha)	Cloth
Letters (kalimah)	Saturday (Sanechar)	Father/Father's Brother	Cloth
Pen (qalam)	Sunday (Itwar)	Father/Father's Brother	Cloth

The chart does not locate the female head of the hearth. The explanation was that she does not weave, she makes quilts. For each specific hearth, knowledge of weaving begins with the head weaver of the hearth and the kumba who, through the pen, receives the craft of weaving and against whose name simultaneously a record is maintained. The transmission of such knowledge moves from the general to the particular, ending finally with hayat.

Beginning with the pen as the constitutive attribute of the first row, we find a grading of each row until we reach the attribute of life. With the exhaustion of the seven rows of bobbins on the beam there is a return to equilibrium. This end, in turn, becomes the starting point of the new cycle and gives time its cyclical form. By locating the cycle in the day and the week a homology is established between everyday production and production for the shroud, between the spiritual and corporeal worlds and between the body of this life and the body of ancestors. In being situated on the beam (in terms of the seven rows of bobbins), equilibrium spatializes the succession of time by substituting for the order of succession the order of simultaneity: it is not only that the material and spiritual worlds, body and soul, are united, but that they are co-present in the work shed.

The progression of time, marked by the beginning of the week with the pen, is counterposed to a time regulated with reference to the computation of the twelve months of the year and the twelve hours of day (light) and night (dark). Similarly by virtue of the correspondence between the two worlds, the action of both living and ancestral bodies on weaving is regulated by reference to the number twelve. On one horizontal row of the wire each of the twelve bobbins has the name of one of the ancestors. In actual reeling these names are not often known. Yet, the belief remains that the names of ancestors are written on bobbins. In this way the present moment (*waqt*) of dying is integrated with the time of the tradition (*zamana*). In contrast to the chronological time of production by the week, the time of the zamana establishes a transhistorical permanence. This time is perpetual: it does not conceal a past which has passed away because its very succession in terms of every death in the community for which cloth for the shroud can be made ensures its return.

Stage Four: Wefting

Here the work shed is sealed off for all hearth members and workers, except the agnatic head of the kumba, who fashions weft members of cloth after he has bathed. While doing this work he recites those du'a mentioned in the *Mufidul Mu'minin* that he remembers. As in everyday production wefting begins immediately after the first prayer call and must end by the last call. After the cloth is made it is left in a heap and not rolled as is the custom in everyday production. In addition, an explicit importance is placed on the number four.

The body of the corpse and the agnatic head of the kumba supply the modality of the signature of wefting. At the same time the loom cumulatively includes the signature of the previous three stages. In warping, warp members combine the quality of dry heat (fire and air) and cold humidity (earth and water). In this union the resultant combination of fire and water, or hot humidity, is found in the body of the corpse, resulting in its upward and swelling motion. Also, in reeling, the corpse is infused with carnality. The point of wefting is to manifest this carnality so that the corpse can enjoy its sojourn in the terrestrial world and is of benevolent disposition (*mizaj*) in its relationship with the world of the living. The visible sign of hot humidity is the colour of the corpse itself. Weavers say that a few hours after death the corpse turns yellowish. This colour imprints the body of the corpse with hot humidity.

The weft members of cloth have the character of air and earth. They must be completely dry before cloth can be made. Second, they must act on the warp members suspended from the warp beam. Like warp members, the weft too is suspended between the feet and head of the weaver at work. But then contact with the loom gives to the weft the quality of coldness since the loom is embedded in earth. The point of contact between the weft and the loom is provided by the shuttle and the comb. While the shuttle is at the level of the loins, the comb is well above the weaver's groin, parallel to his chest.[7] The contact, in other words, is between the upper and lower halves of the body.

In the visible world the union between dry coldness and hot humidity is compared, as the modality of the first two signatures also suggests, to the union between man and woman: the two who compose this combination must become four. In sizing the

---

[7] It may be mentioned that the weft is inserted inside the shuttle and that the comb stitches cloth.

combination between hot dryness and wet coldness indicates the union between bride and groom, while in wefting such unity occurs between the body (*jism*) and spirit (*ruh*) of the person. This leads to a conjugation both between man and woman and the community and its ancestors. The loom combines these four aspects: when unused it has a dormant femininity; when acted upon it is worked by man. Finally, it clothes the body of the deceased by weaving the four elements of nature into cloth. The four elements are represented by weaving any one or more of the four designs.

The specific modality of wefting is just one of the four aspects found in the design and measurement of cloth. Built into the frame of the loom are the three other signatures. The frame is called *tasgara* by weavers, and one way of distinguishing a loom on which cloth for the shroud is made is to refer to it as tasgara or qutb. The two terms are used interchangeably. Qutb refers to a centre or a pole around which the other four pieces of wood revolve, and more generally to the four elements of nature. The pole maintains life in an interior spiritual sense since it is on it that the gaze (*nazr*) of the patron saint Sis Ali rests when he looks upon the wefting of cloth. He is referred to as *katib*, who not only transmits divine instruction, but also maintains an account of the departed. Immediately below his gaze the name of the former agnatic head is inscribed. He is the pole as long as one looks to him for refuge and is referred to as the help or aid (*khidmat*). The third name below the *khidmatgar's* is the pir, the spiritual guide, characterized as the one with patience (*sabr*). In times of distress one should remember the term pir and recognize he is *abdal*, which literally means substitute, in that the physical body substitutes for the person's real presence. As one is recalled to the superior world a member from the rank below takes the recalled person's place. The last name found on the pole is murid, who is the successor to his master. The pole is life which is invulnerable to the peril of the second death, for it has passed the test of the first one, in that it has passed the test of judgment.

Corresponding to the four hierarchically arranged terms are various qualities of colours. The katib dazzles with the brilliance of his white light, while the khidmatgar emits the colour red. The pir typifies yellow and the murid the colour green. In the visible world these colours are manifested by white and black at the two extremes with red and yellow in between. The qutb comprises the four elemental modalities characterized respectively by the four basic terms: water (white),[8] the substance of the body; fire (red), the nature of the body; air (yellow), the image of the body; earth (black), which is the body.

The importance of the number four is strongly signified in the temporality of weaving. I had earlier said that the event does not merely transform into a symbol, but that the event is the person enacting the modality of any one of the four stages. An exegesis of the number four introduces a calendar whose time is measured according to the relative increase or decrease of night and day, while night and day themselves alternate with each other as symbols of the esoteric and exoteric, i.e., the time of the stranger and the daytime of production. Further, a homology is established between the divisions of night and day and those of the ritual calendar which obeys a ternary rhythm. What is revealed ritually to us in this homology is that the act of weaving orders the space of the hearth in which the official ritual of calendar time is transmuted into a ritual of cosmic liturgy, in that the person occupies an existence typified by the modality that rules his or her articulation of the four elements. Such, for instance, is the first division of the day which corresponds with the month of Id, when the ritual year begins for the Muslims of Barabanki. This division must be enacted somewhere by someone. Hence the importance of the zanana and the quilt room that houses the nuptial bed, and of the mardana and the work shed.

While the official ritual calendar commemorates an event, the exegetical calendar (signified by the number four) brings back the event by setting it in the present and by embedding it in categories of people who belong to the hearth and the kumba. In this way the lived situation is projected into the exegetes who, in both everyday life and ritual production, effect the recurrence of the ritual and exegetical calendar. By means of the colour combination and design on cloth such people are the event themselves, and are the time of weaving. That is, the event is not reduced to its causes, but indicates that the recurrence of the

[8] The fact that plain white cloth is the staple production item of everyday life points to irresistible constraints imposed from outside the cosmology of weaving. The nature of these constraints and their implications are not explored here.

daily calendar is also the cycle of the soul's calendar. Because each of the four stages are the recurring presence of the same respective person, they do not merely commemorate an event which happened in the past. The exemplification of each stage as event determines the present since the personal presence of individual figures is projected into the daily calendar.

This event rests upon both an exterior or exoteric cycle of the unveiling of the four stages and an interior occult exegesis. This may be understood as follows: the work of wefting in both everyday and ritual life follows from a period of judgment (qayamat). Through the four weaving stages the visible nature of the world is revealed, followed again by a period of calamity, which succeeds the daytime of production. As qayamat this period is typified through the stranger (ajnabi), spoken of as someone who is not subject to the servitude of either texts or men, and who for this reason is in perpetual exile. Also, since the stranger is not bound by the word his exile is written in his heart. Weavers say that as weavers they are situated between two catastrophes, one of which is the premise of salvation, while the other is the dread of exile, and is perhaps irremediable. This dread is characterized by the 'breaking of the fast'. The person who breaks this fast without service to God is the ajnabi. This destruction of the word, of du'a, zikr, shatters also the symmetry of numbers and so of the visible and invisible worlds, of the above and below, not necessarily in the sense of geometric distance, but ineluctably in the sense of a metaphysical distance. Just as qayamat lends itself to a double meaning, so also the ajnabi is used in two senses. The first refers to the hidden potentialities of destruction which must be renegotiated each day. The other refers to the profanation of the world whose laws of governance are purely mechanical.[9]

THE EVERYDAY AND THE EXTRAORDINARY

In making this cloth weavers explicitly work with the four elements of nature. Depending upon the weaving stage the four

---

[9] This dread is expressed by the weaver in the destruction of the loom initiated by powerlooms and cotton mills. There is a saying among weavers: 'If you want to catch TB then go to Bhiwandi.' Bhiwandi, a centre of the powerloom industry, is the place of exiles.

elements are combined in a specific way. Each combination is infused with a particular modality which articulates a relation between the light and dark. This modality is described by the colour, design and nature of the body. In making it weavers impute to their dead and by extension to their ancestors, an ideal corporeal integrity, which is sought to be emulated in this life. The following chart shows these combinations. (See the combination of the four elements of nature).

THE COMBINATION OF THE FOUR ELEMENTS OF NATURE

Stage	Light	Dark	Colour	Design	Body	Instruments
1 Sizing	fire/air	water/earth	Yellow Green (Black)	◯	Phlegm	Uncured Yarn Lighted Lantern
2 Reeling	earth/air	water/fire	Red Yellow	△	Semen and Milk	Spinning Wheel Yarn
3 Warping	air/fire	earth/water	Green (Black) Red	/\/\/\	Teeth	Bobbins Warp Beam
4 Wefting	air/earth	fire/water	White	↑	Non-material	Warp Threads Loom

An ideal corporeal integrity is given in the significance and measurement of the number four. These social relations also resonate in the everyday act of weaving. In this sense it becomes important to see the link between the everyday and extraordinary not as a division between the sacred and profane, but as being informed by the same cosmology.

In the weaving process the measurement of numbers is not reducible to numbers in statistics, but rests on relationships of equivalence between the past as tradition and the present, and between the body and the work process. In both cases the time of weaving is instituted. In the first case we find the marking of time on the body through the ritual calendar with its numerical divisions and the seven days of the week. This time is spatialized and leads to its periodization. In the second we have the time of the event which lends depth to the numerical points of reference of the calendar and the week. This is the time of the body in relation to the craft of weaving. This is, properly speaking, the time of technique. In the daily cycle this time points to a critical interval where the balance between the light and dark, day and night, given in the number twelve, is broken in favour of either day or night, given in the number four. We find an elaborate correspondence between time as marking and time as activity. What takes place in the interaction between them is not a simple repeated event, but the expression of privileged moments of social activity. Thus, the cardinal moments of both the daily and weekly cycle are placed under a system of cosmic worship.

Following from the above the act of weaving sacralizes on an everyday basis the space of the household and invests with significance the day, the week and the ritual calendar year. Spatially it divides the household into a male and female zone; temporally it divides the day into four units, the week into seven and the year into twelve units. There is, also, an esoteric way in which the space and time of the community is formulated. Here the body of the corpse and the worker providing services are given spatial expression. The division between light and dark, day and night, corresponds to the daytime of production and the time of the stranger. In this chapter I have explored the exoteric and esoteric character of weaving by describing the sequence of each of the four stages (see chart: the sequence of the four stages of weaving).

This chart shows visually what I have attempted to describe discursively in the chapter. The point is to see how the everyday and extraordinary are connected. The links may be explained in the following way. If each category of worker is characterized by a distinctive style, it is important to recognize that each such style is made possible by means of instruments that codify the body gestures of the worker and in this way produce practitioners. The relation between the worker and his instruments creates a field that authorizes social actions. It is because the MB bathes yarn in rice water, then dries the wet yarn by holding it up on two bamboo poles and places lighted lanterns close to the yarn, that

THE SEQUENCE OF THE FOUR STAGES OF WEAVING

		Person	Object to be Worked On	Instruments of Work	Spaces of Work	Techniques of Work
I	Everyday	Bride (dulhan) as the supervisor of children	Yarn	Water, bamboo poles	Compound or boundary at the front of the dwelling	'Education of the feet (pa'e ki ta'lim)', hands, feet. Parda as secluding women, speech as commands.
	Ritual	Mother's Brother (Mama)	Yarn	Rice water, bamboo poles, lighted lantern	Male part of hearth (mardana), platform of the mardana	Hands and feet. Prohibition of sight and speech.
II	Everyday	Woman head of the hearth (walida/ammi) and all other women of the kumba	First reeled bobbin (kunda) and other bobbins	Yarn, two spinning wheels, upright spindle (indi)	Room of the nuptial bed (khatiya ka kamra) within the zanana	Hands, forefinger and thumb. Squatting and sitting posture. Parda as exclusion of marriageable men. Speech as gossip.
	Ritual	FZ (Phupphi) and other non-hearth but kumba women members	Bobbins. No kunda	As above. Yarn must be damp	Quilt room (du'ai kamra in the zanana	As above, except speech. Speech as laments

		Person	Object to be Worked On	Instruments of Work	Spaces of Work	Techniques of Work
	Everyday	Man head of the hearth, his younger brother or son of the man head	Warp beam to make warp members of cloth	Reeled bobbins. Lighted lantern	Workshop (karkhana) in the mardana	Standing fingers, torso, hands for measurement. Speech as commands and as emblematic of the 'inner eye' of weaving
III	Ritual	BS (bhatija) or classificatory bhatija. Always a non-hearth member	Warp beam with seven wires attached to it	Reeled bobbins (always twelve in number). Lighted lantern	Workshed	As above. Prohibition of speech
	Everyday	Man head of the hearth. In his absence his younger brother or eldest son	Loom	Warp members of cloth. Lighted lantern	Workshed	Rhythmic pacing of hands and feet. Always sitting, sharp sight and 'inner eye' of weaving. Speech as recitation of du'a written in the kitab
IV	Ritual	Agnatic head of the kumba	Loom as pole (qutb)	Warp members of cloth. Lighted lantern	Workshed	As above. Speech as repetitive recitation of du'a given in the kitab

The Semiotics of Weaving

he is characterized by a signature. This signature is not available to the bride who cures yarn in everyday life because, as a worker, she can be substituted. Her conjugal household may contract the labour of a member of the community, a contract that is denied when cloth for the shroud is woven.

There are, however, determinate connections between the everyday and extraordinary acts of weaving. Such connections are not limited to the use of the same techniques but also refer to the type of personnel engaged in the various stages of weaving and the coding of space and time. The MB and the bride carry out tasks associated with sizing. Both stand in an affinal relationship with the concerned household. Their work space is fixed in the mardana. While the bride negotiates her status in her movement to the inner part of her conjugal home, the MB dries the yarn in the zanana. He represents an awakened femininity. The dry yarn is reeled into bobbins by the woman head and the other women of the household, or when cloth for the shroud is made, by all the women of the extended family in the zanana of the household. This work is not only collective but also points to the role of the woman as creator and nourisher. Warping is the domain of the younger brother and of the FBS. In the work shed a display of learned skill is a necessity for both the types of workers and rests on an equilibrium of colour and design. Both of them are apprentices. Finally, weft members are fashioned by the male head of the household and of the agnatic line. Work results both in the production of cloth and reproduction of the community.

Conclusion

I have sketched the cosmology of weaving and shown how it informs the everyday and extraordinary life of the weaver. This cosmology is seldom articulated verbally and almost never in the finished discursive form in which it is presented here: as complete and exhaustive. If this tradition is not consciously spoken about, its embodiment in the weaver when he sits down to work at the loom is striking, expressed both in the relationship between his body and the instruments of work and the way the corpse is clothed with ritual cloth. Together, the discourse of the tradition of weaving and the techniques of making cloth show that for the community of weavers the relationship between work and ritual, as well as between the instrumental and symbolic is predicated not on their separation, but on their unity. This is seen in the relation between weaving and the body. By codifying the gestures of the body, techniques fabricate social relationships and in the process delimit the worker's body in work. For this reason, individual workers are transformed into signifiers of rule and the tradition of weaving is incarnated 'as flesh'.

Simultaneously, with the fabrication of cloth for the shroud, the body of the corpse acquires an ideal corporeal and spiritual integrity. The dead body is reinvigorated with carnality and spirituality through the four elements of nature and the four categories of workers.

Finally, I have argued for determinate connections between the everyday and extraordinary acts of weaving. The most obvious link between the two refers to the techniques of work. The connection is also found in the way weaving encodes time. In everyday life the act of weaving divides the day into four parts. Each part has a category of worker and a segment of the ritual calendar associated with it. When cloth for the shroud is made these divisions are formalized. Also, an explicit significance is attributed to the seven days of the week. The esoteric conception of time, faintly discernible in everyday life, achieves its fruition in the extraordinary world of the weavers when the corpse finally becomes available as an ancestor. Through an analysis of the loom as 'pole' (qutb) I have argued that through this notion of time the tradition of weaving is incarnated.

Chapter Four

Work, Worship, Word: A Study of the Loom

The study of material objects as providing insights into the culture of communities flows in one of two directions. Material artifacts are examined either in terms of the use they have for the community in question[1], or their expressive nature is privileged in the analysis.[2] In the first case the semiotic of such objects is reduced to their utility, while in the second we find that in constituting the material thing as a signifying object, its specificity is lost from view. When the two dimensions in the study of the object are combined[3], the analysis often excludes the relation of the worker to his instrument of work. In this chapter, I seek to combine these two aspects by focusing on one material object, the loom. By looking at the loom as it is constituted in the actual practice of producing cloth I have tried to understand work as a unitary phenomena. Of special relevance for me is the fact that the practice of weaving in the community conjoins the act of weaving and the act of praying.

---

[1] See the volume edited by Esther Goody (1982), especially the introduction and the first chapter.

[2] This, for example, is the general thrust of Humphrey's (1971) article on Mongolian spirit figures. See also Messick (1987). Marglin (1990) in his study of the weavers of Orissa is mainly concerned with contrasting non-western notions of work with occidental ones. The latter ostensibly privilege the separation of planning from execution. As far as Oriya weavers are concerned an important dimension missing in his examination is the place of the loom in the life of weavers although an entire analysis is constructed around it.

[3] See Miller (1985), Dilley (1987).

The Mufidul Mu'minin

As material object and significant symbol the loom is classified into its component pieces in the *Mufidul Mu'minin*, a text widely referred to by weavers. The *Mufidul Mu'minin* is a small (twenty-eight pages) primer on weaving written in old style Urdu. In sequence, the first eighteen pages describe the origin of weaving in the world: in the mode of a conversation between Adam and Gabriel (Jabra'il). In this dialogue terms designating the pieces of the loom are introduced in a way that each term corresponds to a supplicatory prayer (du'a). Next each term is broken into its phonetic roots. Further each of the loom's pieces is diagrammatically represented. The last eight pages trace the genealogy of the nurbaf (weavers of light), the term used to depict weavers. Beginning with one of the sons of Adam, Sis Ali, the genealogy concludes with Ayub Ansari who lived in Medina in the time of the Prophet.[4] Weavers take their name from Ayub Ansari and consider Sis Ali to be their patron saint.

The *Mufidul Mu'minin* is a kitab in the sense that Izutsu uses it (Izutsu 1964: 18–24), but with one difference. It is seen not as a revelatory text, but as supplicatory prayer (du'a), in that the text shows how the personal relationship between man and God is institutionalized in the operation of the loom. The du'a mentioned in the *Mufidul Mu'minin* point to a mode of worship

---

[4] During the festival of Chahullam (forty days after Muharram), the Ansaris make elaborate *taziyas* to be ritually interred on the banks of a nearby river. One part of the festival is a ritual of inversion, while in the other the Ansaris reiterate the self-definition of their community. One of the stories in this self-definition describes Ayub Ansari as a khidmatgar. The story unfolds in the following way. Muhammad, on his entry into Medina, was met by a group of distinguished elders, each of whom implored him to grace his house. To offend no one and yet do justice to all the Prophet declared he would accept the hospitality of the person before whose house his camel stopped. The camel, in turn, stopped before the house of Ayub Ansari, a pious but poor weaver. Ayub Ansari gladly received Muhammad but admitted he only had camel's milk and stale bread to offer, and that too only to the Prophet. After Muhammad had washed his own feet he asked Ayub Ansari to feed him. There were, of course unlimited quantities of fresh bread and milk, enough to feed every inhabitant of Medina. Whatever the Biblical points of this story, weavers use it to illustrate their quality as honourable hosts who would rather starve their children than turn a guest away for lack of food.

where sounds (or the phonemes of the du'a) embody virtue. The articulation of these sounds is intrinsic to the act of weaving. Weavers hold that the *Mufidul Mu'minin* delineates the complete meaning of the term Nurbaf. The kitab says the term Nur has three significant letters: Ne for *Nur Haqq* — the light of truth; We for *wahm* — to have no doubt while thinking of God; Re for *raz-o-niyaz* — the secret prayer of God. Such prayer is structured around *wuzu* (ablution), *namaz* (prayer), *ibadat* (worship), *roza* (fasting and abstinence from pleasure), zikr (the invocation of God) and *fikr* (the valorizing of God).

The classification of the loom into its component pieces and the subsequent association of each piece with a prayer in the *Mufidul Mu'minin* is crucial in understanding the relationship between the loom and the community of weavers. Theoretically, the relationship between material object and the community in which it occurs can be modelled on the Saussurian system of an ideal and logical language (Humphrey 1971; Munn 1973). Here binary oppositions are not only organized into an either/or scheme,[5] but are also frozen in time and space. An alternative approach becomes necessary when the frames[6] of such symbols vary with each specific occasion in which they are articulated. That is to say, the relationship between weavers and weaving is better understood if the loom is seen not as a static object but as an artifact in action. In action, then, the loom brings the instruments of weaving, the labouring body and the speaking body into a definite relationship.

There are four distinct 'practices' in which the loom is the central material/symbolic object. These are:
(1) Everyday production of cloth.
(2) Production of cloth for the shroud (kafan).

[5] Humphrey (1971) examines Buryat magical objects within the Saussurian paradigm of analysing langue as distinct from speech. These objects communicate as socialized idiolects, they are non-isologic signs and they stand in a homologous relationship with the myth that is told about them. In each instance there is a neat logical separation: individual/collective, signifier/signified, and sign (language)/object (material).

[6] Miller (1985) uses the term 'frame' as the process by which contextual cues determine the interpretation to be placed on the object. By definition, such interpretation is specific to time and place. As I see it, the frame provides the backdrop for the concerned practice but is not in itself representational. This is consistent with Miller's argument.

(3) Initiation ceremony of young male children into weaving.
(4) Transmission of the loom as inheritance.

The first two practices deal with cloth production, while the latter two show how certain types of people are socialized into the community of weavers. All four practices acknowledge the *Mufidul Mu'minin* in that during them the prayers of the text are recited, either from memory or through reading. In other words, the practice of weaving is informed as much by the physical gesture of the weaver as by his recitation of prayers.

I use 'practice' from the perspective of the male weaver working on the loom. Instead of retaining a sharp distinction between the signifier and signified, Bourdieu's notion of 'practical taxonomies' (Bourdieu 1977: 163–4) as orchestrated practices in the reproduction of the social world appears a more useful way of understanding the four occasions on which the loom comes into use. These practical taxonomies, of which I have identified four, organize the time of the domestic group in particular and the community at large — synchronically and in succession. That is to say, each practice constitutes time in a particular way. The frame within which such constitution occurs is provided by the prayers of the *Mufidul Mu'minin*. I will return later to the prayers. For the present, practical taxonomies, in constituting time, show how weavers reproduce significant divisions of their social world — of age, gender and position in the relations of production. Simultaneously, such taxonomies reproduce these structures, but in a transformed way. The prime instrument of such divisions is the loom.

I have mentioned that these practical taxonomies are framed by the prayers of the *Mufidul Mu'minin*, and in the process the time of the community is constituted. Briefly, in everyday production, prayers, recited once by the head weaver of the hearth, initiate the weaving cycle. Through such recitation the head weaver asserts his right and control over the loom. In production for the shroud, prayers, repeated in a chant by the head weaver of the agnatic line, point to the fulfilment of obligations as well as seal an end in that they are uttered for someone whose death is imminent. In the initiation ceremony prayers are read from the text and then repeated. Such prayers mark a temporal difference and continuity between generations in that they recognize the emergence of one's progeny as working hands. In the

Work, Worship, Word • 119

transmission ceremony prayers are read but not repeated. They establish continuity with the tradition of weaving, as well as acknowledge that one's male progeny will now occupy the position of the transmitter of the loom.

In each of these practices the reciter of prayers stands in a dialogic relationship to the *Mufidul Mu'minin*. It must be remembered, however, that this dialogue is not merely between the reciter of prayers and Sis Ali, the patron saint of the weavers, but is also one where the reciter is in communication with a significant member of his domestic group or his extended kin group who may or may not be present. For this reason, the *Mufidul Mu'minin* is often used to establish the legitimacy of positions, of genealogy, gender and generation, in cases of conflict.[7]

THE MATERIAL OBJECT

Having explained the place of the *Mufidul Mu'minin* in the life of the community as constituted through the act of weaving, I turn now to a description of the loom[8] and its classification. Subsequently, I will relate this classification to the four practical taxonomies enumerated above. In this way we can understand the relation between the material object and the community.

This relationship was brought home to me with dramatic effect. Muhammad Umar, a distinguished head weaver of a particular lineage, had hesitantly consented to teach me the operation of the loom. Before explaining the technical details he gently blew his breath in my ear. He then commanded me to wash my hands and feet. Next, he explained that the distinction between my right and left hand, between my upper and lower body, was the

---

[7] In a recent study of women weavers in a Moroccan town, Messick (1987: 210–55) argues that the subordinate discourse of weaving provides a commentary on the dominant patriarchal ideology by subjecting it to interrogation. This interrogation combines technical and mythical structures in the single act of weaving. The 'said' and the 'craft' are the same things. However, an important component is missing in such a structure. We need to know not only how women constitute themselves as weavers, but also how the material object itself is constructed and thought of.

[8] The *Mufidul Mu'minin* is not dated. Its author in the Persian is Irshadul Muminin, a pseudonym. It is translated from the Persian into Urdu by Maulavi Murtaza Khan. The book was published in Lucknow.

difference between the light and the dark. Such distinctions did not establish a relationship of correspondence between the light and the dark but were characterized by the light and the dark. Just as my body was divided so were the pieces of the loom. 'This is what is meant by nurbaf. We — the loom and I — are nurbaf', he said. Often he would pray while I waṣ learning how to weave. Initially, I thought he was chanting his prayers so that physical damage to his loom was minimized. Later, he explained that prayers are always recited by the head weaver of the household whenever the loom is in operation. Here, every movement of the worker's body is accompanied by a specific prayer so much so that work on the loom is simultaneously word and gesture. This work is founded on the classification of the loom. Thus, I have disaggregated the loom by classifying its pieces according to spatial contrasts, such as up and down, inside and outside, front and back, left and right. This points to Bourdieu's observation: the same set of symbolic relations are organized across different terrains of action through the agency of the habitus. Spatial contrasts follow from the basic division between a light and dark part of the loom affected by the weaver at work. This basic division separates the loom into an upper half (*pankha* or frame) and a lower half (*pavadi* or pit). Other divisions flow from this distinction. The pit is associated with darkness and the frame with the quality of light.

The present loom has fifteen pieces.[9] The pit is composed of two pieces, the *pa'e raftar* or treadles and the *do takhta* or two wooden planks. The frame is divided into twelve pieces which show the spatial division of the loom. One piece (the tasgara or qutb) expresses the quality of both light and dark. Weavers term their loom tasgara or qutb when cloth for the shroud is being woven, and as loom in everyday life. The following chart shows the characteristics of light and dark associated with each of the fifteen pieces of the loom.

[9] The text mentions that the traditional loom had seventeen pieces. Three pieces mentioned in the text are not found on the present loom. These are: a piece of wood fitted on both sides of the grip (the term for this is *sel*), bamboo sticks three yards in length (*karsancha*) and a peg that the weaver had to stamp (*panch chob meekh*). Each of the loom's pieces has an utterance associated with it and each utterance also refers to a particular action of the body. The shuttle has two utterances associated with it. In preparing to work the weaver must recite a particular utterance before he touches the loom. The text mentions nineteen utterances.

The Classification of the Loom according to the Light and Dark Contrast

Term	Light Right	Dark Left	Light Front	Dark Back	Light Up	Dark Down	Light Outside	Dark Inside
1. Khanghi (Comb)	–	–	✓	–	✓	–	–	–
2. Kuch (Reed)	–	–	✓	–	–	–	✓	–
3. Dandiya (Heddle Shaft)	–	–	✓	–	✓	–	–	–
4. Hatta (Grip)	✓	–	✓	–	–	–	–	–
5. Nal (Shuttle)	–	–	✓	–	✓	–	–	–
6. Do Maku (Two Iron Nails)	–	–	✓	–	✓	–	–	–
7. Sutli (Strand of Yarn)	–	✓	–	–	–	–	–	✓
8. Qalbut (Spool)	–	✓	–	–	–	–	–	✓
9. Pavadi ka Lapetan (Breast Beam)	–	–	✓	–	✓	–	–	✓
10. Mekh (Wooden Nails)	–	–	–	–	–	✓	–	✓
11. Do Balna (Two Wooden Bars)	–	–	✓	–	✓	–	–	–
12. Lapetan (Cloth Beam)	–	–	–	✓	–	✓	–	–
13. Pa'e Raftar (Pedals in the Pit)	–	–	–	–	–	✓	–	✓
14. Do Takhta (Two Wooden Planks)	–	–	–	–	–	✓	–	✓
15. Tasgara or Qutb (Loom)	✓	✓	✓	✓	✓	✓	✓	✓

Eight of the loom's fifteen pieces are associated with light, six with darkness and one has the characteristics of both light and dark. This piece is the tasgara, which expresses all the possible spatial arrangements of the loom. The tasgara in the kitab is the name of one of the bamboo sticks of the loom. In the present loom it refers to all the basic poles of the loom. I have grouped the pieces of the loom in the following way:

Group I	comb, reed, heddle shaft and grip. These pieces are materially linked to each other and associated with light.
Group II	shuttle, two iron nails, thread and spool. These pieces too are materially linked to each other. The first two are associated with light and the latter two have the characteristics of darkness.
Group III	breast beam, wooden nails, two wooden bars and cloth beam. The light and dark alternate in this group.
Group IV	treadles and two wooden planks. Both these pieces in the pit are associated with darkness.

In terms of the four groups the relationship between the light and dark is symmetrical. Group I is distinguished from group IV as an absolute demarcation between the light and dark, corresponding to the separation between the upper and lower portions of the loom. Group I classifies the front of the loom, particularly those pieces above the eye-level of the weaver, placed to the right of him, while group IV classifies the invisible part of the loom, well below the weaver's loins. In groups II and III the light and dark contrast is equally balanced, both internally and in relation to each other. The light and dark relationship is proportional. Group II effects a basic division between the front and outside of the loom, and the inside and left of the loom. In group III the four pieces follow each other as a succession of the light and dark. The light part, in both groups, classifies the front and horizontal members of the loom above the eye level of the weaver, while the dark part classifies the instruments either at the back of the weaver or inside other instruments. The last term in the chart, tasgara, is a summation of the pieces of the four groups.

Utterances Associated with the Terms of the Loom

The *Mufidul Mu'minin* mentions that nineteen supplicatory prayers (du'a) are to be uttered in the process of weaving. These prayers are answers to questions that Adam asks Jabra'il when the first loom is being introduced from heaven (*jannat*). The book recounts the following story which I will translate in my own words. The story was read out to me on four different occasions by male weavers and narrated in a group.[10] Women were prohibited from having the story read out to them. In each reading the basic structure of the story remained intact but there were spoken variations of the written word. The rendering and legitimacy of variations depended upon the status of the weaver narrating them. On one occasion Haji Ghulam Nabi and

---

[10] On two occasions I had explicitly asked that the story be read out to me. The text came to my notice when Muhammad Umar mentioned its existence. On questioning him further he agreed to read it out for me. At an impromptu session Muhammad Umar, two of his sons and his grandson gathered in the work shed where Umar read out the *Mufidul Mu'minin*. I was asked by Umar, as were the others, to cover my head during the reading. Subsequently, Umar kissed the kitab, folded it in a piece of cloth and replaced it in the karkhana. The second reading was more public. With the intention of finding how widespread was knowledge of the text I approached Ghulam Nabi and asked him whether he had the *Mufidul Mu'minin*. Ghulam Nabi said he not only had the text but that he could recite almost all of it from memory. After being prodded he agreed to read it out. He drew up three cots outside the courtyard of his house and asked his sons to spread word that he was going to read the *Mufidul Mu'minin*. Soon almost the entire area aroung Ghulam Nabi was peopled by young male weavers. The reason for this became apparent as soon as he commenced the reading. Unlike the hesitant Umar, Ghulam Nabi was fluent and spoke with great force. Often he would pause for dramatic effect and intersperse his reading with the conviction that the Ansaris would redeem the promise of the *Mufidul Mu'minin*. During the reading some weavers began to cry and one of them loudly lamented that the Ansaris had moved very far from the ideals of the text. The third and fourth readings were tame occasions where Sadiq Ali had the text read out in his work shed, ostensibly he maintained for my edification, but I suspect also for his. The reading was undertaken by one of Sadiq Ali's friends in the presence of three other male weavers. These two readings were not complete since the reader could not fully decipher some of the words.

Muhammad Umar, both head weavers, provided contrasting accounts of a detail of the story. The Haji evoked his *Haj* to legitimate his story, while Umar, more than seventy-five years old, maintained it was his paternal grandfather who told him the detail he was presently narrating.

The story opens after Adam is expelled from Jannat for eating wheat. Adam feels hungry and prays to Allah for food. Allah orders Jabra'il to give Adam a piece of wood named Salim, and a goat and to teach him to cultivate food. Adam and Hawwa (Eve) slaughter the goat and eat its meat. Adam next complains of his nudity. (Some weavers say Hawwa taunts Adam of his nudity, while others argue a houri, not Hawwa, offers to marry Adam if only he can clothe himself). Allah orders Jabra'il to give Adam a box full of weaving implements and to teach him the craft of weaving. (Opinion was divided on what this box was made of. Ghulam Nabi emphatically argued the box was made of the wood named Salim and that the present day *gathuwa*[11] was originally placed inside such wood. Muhammad Umar insisted that pieces of the loom were fashioned from this wood). Jabra'il lets Adam know Allah has sent a box of weaving implements for him. Adam offers prayers, opens the box and finds the seventeen pieces of wood. I have classified these into the four groups.

Adam then asks Jabra'il to teach him to weave. Jabra'il says the prayers to be uttered in the process of weaving are equivalent to reciting the Quran one thousand times, or feeding two thousand needy people and setting free one thousand camels in God's name. In retaining prayers in his memory the weaver is protected from calamity. If, however, he practices his craft without reciting prayers and continues to call himself *Momin* (faithful Muslim) he is a liar, barred from entry into the Muslim community on the day of judgment. Adam finally asks Jabra'il questions concerning each piece of the loom.[12] Jabra'il orders Adam to gird up his loins.

---

[11] A gathuwa is a bundle in which three pairs of heddles and combs are wrapped, to be used on ritual occasions.

[12] Lack of space does not allow me to mention all these prayers. Suffice it to say that each piece has a prayer associated with it. The fifth, sixth, seventh, fourteenth and sixteenth prayers are not recited at all.

(1)	Adam:	While girding up the loins which prayer should be recited?
	Jabra'il:	*Allahu Akbar, Allahu Akbar, ashhadu-an-la Ilha ill-Allah wa-lillah al-hamd.*
(2)	A:	While tossing the shuttle with the right hand which prayer should be recited?
	J:	*La ilaha illa 'llah.*
(3)	A:	While tossing the shuttle with the left hand which prayer is to be recited?
	J:	*Muhammadun rasulu 'l-lah.*
(4)	A:	What prayer is to be recited at the time of working the treadles in the pit?
	J:	*Ya hayiyun ya qaiyum ba rahmat astaghfiru 'llah.*
(8)	A:	While holding the thread which prayer is to be recited?
	J:	*Ya ma'budan ya maqsud.*
(9)	A:	While holding the reed which prayer is to be recited?
	J:	*La ilaha illa 'llah-al-Malik-al-Haqq al-mabni.*
(17)	A:	And while holding the spool of thread?
	J:	*Al-hamd li Allahe Rabb al-alamin.*
(18)	A:	While holding the heddle shafts which prayer is to be recited?
	J:	*Inna Rabb-i ala Siratu 'l-mustaqim.*
(19)	A:	And while holding the comb?
	J:	*La ilaha illa 'llah wahdahu la sharik lahu.*

These prayers, transliterated from the text, are remembered and recited most frequently. Most weavers I talked with could recall the excerpts I have presented. Muhammad Umar remembered the fifth, sixth and seventh prayers in addition to those cited above and Haji Ghulam Nabi could recite almost the entire text. Both these weavers remember prayers associated with those parts of the loom not found in the present one. Muhammad Umar, Ghulam Nabi and most other weavers believe that the prayers of the text consecrate the act of weaving, and it is only through their recitation that Ansaris establish the nobility of their craft. This is done by constituting a formal discourse centered around the self-definition of Ansaris as *momin*.

The recitation of prayers, whether memorized or read, whether repeated or enunciated once, serves to distinguish one practical taxonomy from another. In effect, verbal styles of enunciation establish frames of each of the four practical taxonomies. These prayers are of a kind where the personal relationship between man and God is institutionalized through the material artifact. This is evident in the way prayers, in framing practical taxonomies, structure the time of weaving and the domestic group. Before describing the process by which time is structured I will show how the loom is constructed in language.

Utterances as Worship

The prayers listed in the *Mufidul Mu'minin* are a type of worship understood as du'a. Du'a, described as personal prayer addressed to God, is distinguished from *salat*, which is ritual or liturgical prayer. (Gardet 1965: 617–18). When addressed to the common good of the community du'a assumes a recognizable ritual form: it uses the procedures for salat. In uttering a divine name repetitively du'a links its request with the evocation of each name to each attribute. In weaving du'a cannot be substituted by any other form of prayer since the phonic sounds of du'a encapsulate an idealized past in that they address questions to an ideal figure. To the extent that utterances associated with the loom evoke divine names in repetition, du'a are appeals trusting in divine mercy.

The *Mufidul Mu'minin* maps out the semantic domain of du'a. The text mentions that after Adam has learned the prayers Jabra'il tells him that each of them has associated ideas. Adam wishes to know these ideas so that he can teach them to his children. What follows is the same question/answer format that was used for the du'a. The ideas associated with prayers proceed by breaking the phonetic root of the terms linked with the loom's pieces. These ideas cut across the four groups. I will mention some of them.

(1) The term *khanghi* (comb) has five letters:
    'Ka' for *kinaya*   – to remove ill feeling
    'Ha' for *hawa*   – to remove greed
    'Na' for *nasihat*   – to learn religion from elders
    'Ka' for *kufr*   – to remove blasphemy
    'Ya' for *yari*   – to avoid attachment with worldly things

(2) *Hatta* (grip) has three letters:
    'Ha' for *hidayat*   – to become righteous

'Ta' for *ta'iba* — to repent for misdeeds
'Ya' for *yad* — to remember the bad deeds of the self

(3) *Nal* (shuttle) has three letters:
'Na' for *nafi* — to negate/deny (an opulent life)
'A' for *asbat* — to affirm the existence of God
'La' for *latif* — to be pure

(4) *Sutli* (yarn) has five letters:
'Sa' for *sitam* — to resist tyranny
'Wa' for *wajib* — to understand as necessary deeds declared true for a human being
'Ta' for *takabbur* — to keep away from pride
'La' for *lutf* — to have mercy on God's creation
'Ya' for *yari* — to establish friendship with religious deeds

(5) *Balna* (wooden bar) has four letters:
'Ba' for *bari ta'ala* — to remember God
'Ya' for *yamin* — to observe fasting during the day
'La' for *li'l-lah* — to awaken at night with the name of God
'Na' for *nafs* — to consider carnality as mortal

(6) *Takhta* (plank) has four letters:
'Ta' for *tartib* — to learn the mystic way sequentially
'Kh' for *khatra* — to be unafraid of others
'Ta' for *tarik* — to lead the life of a recluse
'Ya' for yari — to befriend the poor

These prayers sketch out a moral universe in which sounds embody virtue. Together with the semantic domain of the pieces of the loom they constitute the material object. They show that if the loom is embedded in its instrumentality, in that it demarcates technical terms, it is simultaneously a construct of language to the extent that these terms are indissolubly linked to specific prayers. In uttering them the weaver invokes the loom as memory. Du'a in this sense function as mnemonics in establishing a continuity between the patron saint and the contemporary weaver. This manner of using du'a is theoretically the 'magical' use of language[13]

[13] From a different but related perspective Dilley (1987) has argued that the Tukolor weavers of Senegal associate weaving with spiritual forces.

(Izutsu 1964: 196). The loom in this sense is a construct of the magical word, best understood as: to weave is to pray, but equally, to pray is to weave.

In sketching out a moral universe these prayers structure both the time of weaving and of the domestic group. In everyday production du'a mark the commencement of work on the loom, as well as the category of the weaver associated with this period. In weaving for the shroud and the initiation ceremony utterances are repeated, and readers and reciters of them are using du'a, especially in the sense of metrical repetition, to evoke divine names and associate their attributes with the relevant pieces of the loom. The transmission ceremony does not involve the operation of weaving, but prayers associated with it show the place of the loom in the community.

The production of cloth, both in everyday life and for the shroud, establishes the temporal boundaries of weaving by the week, while in the initiation and transmission ceremonies the loom restates the relationship between man and God as much as it structures the boundaries within which work in the domestic domain is constituted. These boundaries are structured on the basis of those who work the loom (men) and those who will provide services to weaving (women). The initiation ceremony separates men from women, while the transmission of the loom differentiates men into generations and affirms the community of weaving from father to son. To see how time is structured in this way I will discuss each of the four practical taxonomies.

The Dialogue of Weaving

Thus far I have argued that the loom as an object of study has fluid boundaries, ranging from three or four sets of utterances

---

Weavers are considered both by themselves and by others to be experts in magic for weaving is a magical act. The magical notion of weaving is located in the construction of the loom itself. In turn, such construction is linked to the performance and expression of the ritual space of weaving. The plan and construction is both a physical representation and expression of the weavers' system of ideas, thereby showing that the loom is located as much in the symbolic as in the instrumental world. The emphasis of this study is different from Dilley's article. The loom constitutes the practice of weaving as both dialogue and physical activity in simultaneity.

(everyday production) to utterances comprising almost the entire text of the *Mufidul Mu'minin* (the transmission ceremony). More important, we have to determine whether these utterances are absolute and unconditional. If they are considered the utterances of God they must function independently of the circumstances that evoked them. Their meaning is not qualified by an audience whose potential reactions have to be taken into account. Contrarily, I will argue that in their usage these utterances are not absolute but establish a dialogue between man and God and man and man. Each practice establishes a specific relation between the addresser and addressee located within the life cycle of the domestic group.

The time of weaving, instantiated through a dialogue between the addresser and addressee, can be made evident if, in each practice the addressee is specified. At the outset the patron saint of the weavers is the addressee of the four practices. He makes available the tradition of weaving by supplying present day weavers with a past that has been divinely ordained, not as revelation but as supplication. The particularity of the addressee becomes available when we situate utterances within the hearth and the agnatic line. Accordingly, I will discuss each of the four practices of weaving as dialogic communication.

Everyday Production

It takes approximately seven days, from the time yarn is cured till it is woven into cloth, to exhaust one cycle of cloth production. The loom comes into use on the second or third day of the cycle. The first three prayers and occasionally the fourth, are recited by the male head of the hearth working on the loom. Though said once, the weaver, in reciting such prayers reiterates his status as the head of his household. His younger brother or son/s will not pronounce these utterances. In the head weaver's absence utterances are not said for everyday production. Weavers maintain it is not necessary to utter du'a more than once in the weaving cycle, but these appeals must be made as soon as the weaver begins his quota of production for the week. In this sense du'a mark the initiation of each cycle. Prayers must be uttered after the morning call, when the weaver begins work.

To recapitulate from the second chapter, a complete weaving household is composed of three generations with an ancestral loom. Ideally such households are composed of four categories of workers engaged in the weekly cycle. The bride cures the yarn while the woman head reels it into bobbins. The eldest son fashions the warp, and the male head the weft. There is considerable interchange of tasks between the first and the last two categories, but men do not do the work of women. Women on their part are prohibited from fashioning the weft. In a two generation household the woman head often does the warping, but only because this type of household is in the process of negotiating its status as an independent hearth. Rivalry and competition between families living in the same household often crystallize around the loom as property.

In everyday life the head weaver's right to recite prayers is contested most often when one of the male members of the upper generation is in the process of establishing his own household. An incident that came to my notice concerned two brothers, Sadiq Ali and Ashraf Ahmad who belonged to a two generation household. (See chapter two for a composition of the household). As is the custom, the household would split once the members of the lower generation had reached the age of marriage, provided the two brothers had offspring who could intermarry. Ashraf Ahmad operated the inherited loom and his younger brother used but did not own the second. In establishing his own dwelling, Sadiq Ali insisted Ashraf part with one of the two looms. When Ashraf refused, the younger brother began to recite prayers while working on the second loom and in this way staked his claim over it. The two brothers resolved the conflict by allowing Sadiq Ali control over the loom. In return Ashraf would not fully finance the construction of the new dwelling. In a later conversation with me, Ashraf said he had parted with the loom because it bore the imprint of his brother's word.

In the above example, Sadiq Ali used prayers in a tactical sense, but in a way that Ashraf was compelled to recognize their usage. This is because du'a serve as ritual markers of the weekly cycle by showing how the work tradition interacts with the specific occasion of making cloth. In effect, the instrumental experience of weaving and the phonetic quality of utterances intermesh with each other to signify a beginning. This beginning is made when the male head sits down at his loom. To this extent utterances are formalized. The absent addressee is evoked once by the head weaver. When asked why such utterances could be recited only by this weaver I was told, 'It is his right (*haq*)'. This right devolves upon him by virtue of his position in the household, suggesting that he is the legitimate transmitter of the tradition. Because this position is singular it cannot be occupied by more than one incumbent simultaneously.

The time of everyday production, posited as a beginning and as a right, is oriented toward the future — a time when collective labour concerns itself with meeting production requirements. In the estimation of weavers this beginning is a mark dividing the household into two ideally reciprocal units — those who work the loom and those who supply their labour to it. When the beginning is mooted as a right it recognizes that the position of the head weaver is open to competition. Within the same generation competition indicates property conflicts and across generations it signifies a time of flux. In its widest sense this beginning points to a period of productive growth which must multiply and increase the quantity of valuable things and people.

In reciting these utterances the weaver as author personifies reported speech. He emphasizes the typicality of the marked utterances. This is made still more evident in cloth production for the shroud, characterized as it is by the repetition of utterances. In everyday production the right of the head weaver to recite such utterances, especially in a two generation hearth, may be interpreted as his and his generation's claim to establish control over the loom and its produce. The addressee is a potential competitor. He is either the head weaver's younger brother or his son. In both cases it is implicitly recognized that this right is transitory, to cease as soon as the weaver retires from everyday production. He retires from everyday production when the loom is transmitted to the next generation.

The temporality of utterances also works in a different sense. If the recitation of prayers structures the week, through them the weaver evokes his work tradition by using prayers as mnemonics. Utterances, used as aids to memory, mark the continuity of the work tradition. This will be discussed again when we consider the first two practices together, since in making cloth for the shroud utterances are both memorized and repeated.

Cloth Production for the Shroud

In the second type of practice ideally all the prayers should be recited by the weaver working the loom. Each utterance is repeated a number of times. Muhammad Umar says that most weavers are unable to repeat all the du'a linked with the pieces of the loom. Consequently, the selection of utterances depends on the memory of the weaver. Most often weavers repeat prayers associated with significant pieces of the loom. Such pieces are those which, in their operation, involve the direct physical effort of the weaver, as well as those that wear out frequently.

When cloth for the shroud is made workers do not belong to the household of the member for the whom the cloth is made. The mother's brother or wife's brother cures the yarn. He carries out this task in his capacity as an affine. The yarn is collectively reeled by the father's sisters or husband's sisters and all other women of the agnatic line provided they are not members of the same household as the member for whom the cloth is being woven. The warp is fashioned by the father's brother's son. The weft members are made by the agnatic head of the lineage.

While in everyday production the addressee is a potential competitor who is not categorically emphasized, in production for the shroud the addressee is a marked person, someone for whom cloth is woven and who, for this reason, can never be a competitor. Unlike everyday weaving, work on this occasion is not thought of as a right, but as an obligation every member of the community is required to fulfil.

Viewed as activities in their own right, the first two practices are opposites of each other. The opposition between the two is, in the estimation of weavers, homologous to that between the light and dark, day and night.[14] The first practice, initiated with sunrise and the first call, marks a beginning and is oriented to the future, while the second, initiated after sunset and the last call, seals an end by establishing temporal breaks in production. Seen together, they establish a continuity with the tradition of weaving. In both practices memory operates as the evocation of the work

---

[14] I have explored the significance of day and night and light and dark in the previous chapter.

tradition by the individual weaver. Memory accomplishes a double task in one act. It represents the past while understanding it to be an idealized practice of the material world. That is to say, only if the past is removed to a distance is memory able to enter into a free relation with the present act of weaving. It was common for me to hear comments by weavers to the effect: 'Since the time of a *zamana* the tradition of weaving has been continuing'. Zamana, as used by weavers, is the most general term denoting time. In this specific case it may be translated as longue duree, referring as it does, to the introduction of weaving in the world. Zamana is opposed to waqt, which usually refers to a particular moment, for example, 'within this time — waqt — we can not mount bobbins'.

This zamana is written in the kitab but its meaning depends on specific practices with which utterances are associated. In the first two practices weavers make the past their own by bringing it into a positive relation with the present. Through its assimilation into the present this past is legitimated. The affirmative act of memory integrates the zamana into a free sense of the present moment. Whenever this integration is not fulfiled the present act of weaving and the work tradition necessarily conflict with each other.

One such conflict that came to my notice concerned a young, recently widowed woman with two daughters, one of whom was at the suckling stage. The woman and her husband had recently established their hearth with an independent loom. Within six months of this event her husband was killed in a road accident. Abandoned by her husband's father's household, the widow appealed to her affinal relatives to help her tide over the inevitable financial crisis. Her husband's elder brother, taking offense at this humiliation,[15] demanded that the loom be placed as collateral with him before any money could be loaned. The widow refused. Instead she started to work the loom herself, complaining of her HeB's (husband's elder brother's) callousness. Weavers of the village complained to the weaver's panchayat, pleading that the widow be forced to surrender the loom to her HeB, since she

---

[15] He felt humiliated because she had appealed to her distant relatives and not to her husband's immediate consanguines. The husband's elder brother maintained that by her actions he had lost honour. I could not get the comments of the widow but her younger brother told me that she could expect no sympathy from the members of her conjugal family.

was disqualified by her sex from making cloth for the shroud. This disqualification was doubly accentuated: the weaver was not only a woman but she also did not know how to recite the du'a associated with weaving, though she knew what such prayers were. Male objections raised the fear that she would ritually violate the loom. She was forced back to her natal household. Zamana, in other words, cannot be challenged. Here, the repetition of utterances integrate the zamana into waqt neither as cycle nor as essence, but as answers to specific questions asked of an ideal figure.

Viewed together, everyday and ritual weaving are finely intermeshed in commemorating a work tradition. But an inner boundary lies between them: utterances associated with kafan cloth are not the ordinary utterances recited every week by the head weaver of the hearth. Within the limits of an abstract calendric unity the concrete time of human life is subdivided into the four segments of weaving. That is to say, the motif of death is transformed and renegotiated through the temporally sealed off sequence of the four stages of weaving (sizing, reeling, warping and fashioning the weft). Within the hearth where death occurs its motif takes on the meaning of a finality. In this sense the addressee of the second practice can never be a competitor: the link with fertility is disrupted.

Yet this motif exceeds calender time since the type of work involved in making cloth for the shroud establishes a living relation with the community of weavers — a relation based on collective labour. The inner boundary becomes sharp and precise in the style and repetition of utterances. In reciting them the weaver does not claim identity with the patron saint, but restates and repeats the avowed vocation of weaving, interpreted in this practice as the obligation to make cloth for the dying.

INITIATION CEREMONY OF MALE CHILDREN INTO WEAVING

In the month of Bade Pir all the utterances are recited in their sequential order and then repeated. The head weaver, in whose household initiation occurs, reads out Adam's questions and Jabra'il's answers from the kitab during the first six days of the month when both the loom and the work shed are ritually cleaned (*karkhana ki safai*). It is imperative that prayers be read during daylight since weaving is rigidly ordered along the light/dark contrast during initiation. The ritual cleaning of the work shed involves, first, a dismantling of the loom and a reflooring of the shed, and second, the loom's overhauling and refixing. The reading from the text lasts for the duration of cleaning and occurs in the shed. Only after this room is plastered over with beaten earth and the loom reassembled, does the initiate attend a ceremony, held in the seventh day of the month, in the village mosque. After this ceremony he is formally eligible to work the loom though he cannot recite du'a.

The work shed is divided into two types of activities. First, bobbins are mounted on the warp beam and warp members of cloth are fashioned by drawing thread from the beam to the loom. Second, weft members are formed through the operation of the loom. The work shed is the place, par excellence, of the male of the hearth during everyday weaving, and of the agnatic line when cloth for the shroud is made. The work shed is closed to the outsider, who may be, depending on the occasion, an outsider to the hearth or the agnatic line. When cloth for the shroud is made the work shed becomes a hermetically sealed unit to the extent that not even the head of the hearth is allowed entry.

One important feature of the initiation ceremony is the emergence of progeny as working hands who will both contend with the everyday production of cloth and inherit the tradition of weaving. Utterances recited here recognize this emergence because they are directed at the patron saint on behalf of the initiate. After the ceremony he is formally segregated from his female sibs.

When in the initiation ceremony the head weaver substitutes for his novice child, the categorical boundary between the addresser and addressee becomes blurred. The problem is of separating one voice from another. This separation is inadequately expressed if we merely locate the type of person reading utterances. Crucially, the style of reading marks the separation between addresser and addressee. When I inquired into the necessity of the head weaver reading utterances I was told: 'his [the initiate's] time (waqt) has not come'. Temporal separation is established through utterances which are read and repeated in a modulated voice. The reader as author of reported speech recognizes that the initiate will take the head weaver's place. In this sense progeny emerges as a value. The value is of initiating the novice as nurbaf.

The substitution of the addresser by the addressee points to the notion of reading and repetition as subjunctive. Implicit in this reading is the positioning of the addresser as addressee. For this reason such subjunctivity questions the fixity of the head weaver's position. That is to say, the temporal separation this reading affects is premised on positional fixity. This position establishes an uninterrupted link between the life of generations and a strictly delimited locale: the work shed and the village mosque. Through prayers this link replicates the relationship of time to space in that the work shed and the mosque together are thought of as the ideal locus for the entire life process of the male weaver. Weavers often express the view that the mosque and the work shed are interchangeable spaces in the work of weaving. A weaver remarked to me that on his son's initiation ceremony into weaving he did not offer prayers in the mosque for the first six days of Bade Pir since he was praying in the work shed.

The temporality telescoped in the initiation ceremony points to the work of weaving as cyclical repetition. This interpretation is suggested since the weaver personifies authorial speech when he reads from the kitab. Reading and rereading, however, operate simultaneously in two directions: first the basic communicative relationship established between man and God provides for the view of the zamana as open: the present head weaver will be integrated into it. He, as much as the initiate, negotiates his status anew. Second, such prayers make available the conditions for constituting one's progeny as productive hands and reproductive males.

In the latter sense the initiation ceremony constitutes time as a stage in the development of the domestic group. This development elevates the ideal of human life by referring to a tradition whose genesis can never be actualized, but which must be negotiated after the birth of every male offspring. With the initiation of every male child it becomes necessary to make contact with this tradition, but at a new stage of development. What is retained is the inner aspect of the movement of an individual life from one developmental stage to another. This interiority is concretely marked within the work shed as a place of worship. In effect, the initiation ceremony points to a transitional stage in the life of both the initiate and the domestic group. Time is spatialized and negotiable. It reaches its culmination in the last practice.

The Transmission of the Loom as Inheritance

When the loom is transmitted from a senior to a junior generation the entire conversation between Adam and Jabra'il is read out once by a holy man. The transmission of the loom is from father to son. In cases where the father is deceased the father's brother or the head of the agnatic line substitutes for the father. The transmission is witnessed and consecrated by a holy man, preferably a Haji. The ceremony occurs from within the work shed of the transmitter's dwelling. The transmitter and the inheritor sit on one prayer mat with their heads covered while a holy man sits on another. Before the reading begins the reader takes the name of both father and son. They are the addressees and the holy man the addresser. After washing his hands and feet the holy man commences to read the first eighteen pages of the *Mufidul Mu'minin*. He will read this section once. Most weavers insist the holy man explain the story to them. In interpreting the story he often embellishes the account of the kitab with his own personal experiences in Mecca. Even in the case of a linear reading of the text reported speech infiltrates authorial speech with commentary.

The reading of the text, known as *nazrana*, refers to the gift of the loom the father makes to his son. In turn, the son makes a weekly gift of food and cloth to his father's hearth after establishing his own dwelling. If the son remains in his father's hearth he holds a feast for the lineage after the inheritance. If the inheritor is the youngest son then his elder brother's forsaking of his right over the loom is called *qurbani* (sacrifice).

Food is sent regularly while cloth is gifted once. Opinion is divided on what such cloth points to. Some weavers maintain that in weaving cloth the inheritor acknowledges that this loom can make cloth for the shroud and is the property of the agnatic line, while Ghulam Nabi told me that this cloth symbolized the shawl Adam wove out of goat's skin after Jabra'il had taught him to weave. After transmission the loom becomes ancestral (*khandani*) property. A loom is also considered ancestral if, prior to its transmission, cloth for the shroud has already been woven on it.

As weaver the Ansari is outside the stylistic composition of the utterances. But it is here that he stands in a sharply dialogic relationship with the prayers of the transmission ceremony. Such utterances mark the weaver as participating in a past which appears as a represented event, itself fixed in the kitab. The represented event is not identical with the processes of weaving, if only because the author of this event is first a holy man, and then a weaver.

What then is the relationship between the represented event and the transmission of the loom? This event marks a temporal break within the domestic group. If utterances of the initiation ceremony distinguish between those who provide their services to the loom and those who work on it, those of the transmission ceremony complete this process of differentiation by recognizing the right of the lower generation to be master of the loom. More important, the representations of weaving in the kitab establish the boundaries within which the practices of weaving institute and periodize their temporalities. Representation in such practices is the emergence of the dialogical other. To weave cloth is to put these representations into motion. The transmission ceremony establishes a continuity with the initiation ceremony to the extent that it finalizes and recognizes the temporal break between generations. But it also exceeds this break by acknowledging that the representations of weaving in the kitab provide a frame through which the work practices are negotiated.

Summary and Conclusion

Everyday and ritual manufacture authorize a temporal continuity by privileging a memory found in prayers while the initiation and transmission ceremonies establish contact with the work tradition by situating this memory in a determinate place. In the initiation ceremony this location is in the domestic domain and in the transmission ceremony it is localized in the work shed. Through this notion of recall we find the emergence of progeny as working hands and the recognition that the status of a head weaver is transitory. To this extent, the work tradition as memory is fashioned by external circumstances. This memory is regulated by the act of alteration not only because it is marked by external occurrences, but also because it is recalled in every act of initiation and transmission. Because of the capacity of alteration the work of weaving cannot be thought of as unitary.

When I say the utterances of the *Mufidul Mu'minin* have a temporal dimension built into them I mean that the boundary between the actual world as source of representation and the world represented in the kitab is not absolute and impermeable. There is a continuous mutual interaction between them. This interaction is ordered through a temporal continuity with the tradition, on the one hand, and marks temporal breaks in the life of the weaver, on the other. Here, notions of time are evoked variously as memory, succession and repetition. In its widest sense the temporality of weaving is constituted through the addressee or the other.

Seen from within the perspective of the four practices of weaving dialogue is action in which the transmission and reception of utterances occurs simultaneously, not sequentially since it is the product of a social situation in which real and potential audiences, earlier and possible later utterances all interact (Volosinov 1973: 72). In reciting these utterances for his work the weaver participates in an already constituted work tradition, but also uses this position to establish and reemphasize his place in the domestic group and the community of weavers. Such consolidation operates almost invariably in relation to a specific addressee. Through his work tradition he addresses his patron saint while through the specific practice he addresses a member of either his household or his agnatic line.

Further, if by his work tradition the world of weaving has been formally constituted for him, specific practices establish the work relationship between a particular addresser and addressee. Here, utterances can only be used in the form in which they are written. Depending on whether the weaver recites them once or repeats them from his memory, or reads them out from the kitab, both the authorship and style of reciting utterances changes. The style of recitation recognizes the presence of the other to whom the speaker is speaking and simultaneously acts as a framing device in three ways.

The frame governing the tradition of weaving is provided by

a narrative that occurs as a conversation. This conversation determines the form of the utterance, the sequence in which it will appear, and crucially, the part of the loom to which it is connected. Second, utterances operate in two ways: when read from the kitab both the questions and answers are recited, and when repeated from the memory of individual weavers only Jabra'il's answers are recited. Third, besides expressing a work tradition these utterances establish and mark a difference between men (weavers) and women (workers). This mark establishes the verbal identity of the weaver and prohibits non-weavers from working the loom. In effect, the recitation of du'a gives validity to each of the four practices and distinguishes one practice from another.

These prayers are composed of a 'theme' and 'meaning'. Thematic unity in the prayers of weaving cannot be reduced to a sentence. For example, the second utterance, 'La ilaha illa'llah', may be broken into its syntactic structure, but it becomes an utterance when said in a specific time and place. Only then does it acquire a theme which, because it includes the specificity of time and place is characterized by uniqueness and indivisibility.

The uniqueness of the theme of an utterance is opposed to its meaning, which is reproducible and self-identical in all in stances of repetition. Meaning can be understood through its morphological and syntactic structure (Volosinov 1973: 100). In the context of weaving the meaning of an utterance is divisible into four distinct practices, each of which is a technical apparatus for implementing the theme of weaving. The technical means of articulating a particular consciousness is a matter of style. Style, here, is the phonic mode within which weaving operates, and is linked both to the discourse of a particular generation at work on the loom, and the mode of prayer in the mosque, with which it is often contrasted. For example, the second utterance, 'La ilaha illa'llah', when recited during everyday production has the stylistic quality of being flat and without any inflection in the voice of the weaver. In weaving for the shroud its repetition is identical with the chant of the evening prayer. In the initiation ceremony the weaver reads it out from the kitab and then repeats the reading. The reading is modulated. In the transmission ceremony, the holy man reads out all the utterances in a flat voice.

Recitation from memory (i.e., during everyday production and for the shroud) legitimates the type of cloth that is to be made, while reading aloud in the initiation and transmission ceremonies marks an entry into the work tradition. Together reading and memorized repetition codify both the work tradition and cloth production. In other words, reading aloud and recitation encapsulate the cultural practices of cloth production. Volosinov identifies two stylistic devices that separate one voice from another in 'authorial' and 'reported' speech: linear and pictorial (1973: 115–23). The linear style maintains the integrity of the speech being reported, while the pictorial style infiltrates reported speech with authorial retort and commentary. Through the first meaning is broken into its referential units. With the second the utterance is put into quotation marks. In this dialectic between theme and meaning weaving becomes dialogic.

The distinction between authorial and reported speech locates the speaker in the practices of weaving by separating one voice from another in the dialogue. In weaving authorial speech corresponds to the transmission ceremony, when the kitab is read, while reported speech is linked to repetitive (cloth production for the shroud and the initiation ceremony) and memorized (everyday production and for the shroud) enunciation. The initiation ceremony combines both authorial and reported speech. When the weaver reads utterances from the kitab, especially in the transmission ceremony, he points to weaving as ordered, sequential and depersonalized. This arrangement contrasts with the utterances of the first three practices. In cloth production for the shroud and the initiation ceremony, as distinct from everyday weaving and the transmission ceremony, utterances are metrically repeated. Here the boundaries of authorial speech are ruptured. This repetition is either memorized (the first and second practices) or read (the third and fourth practices). Through memory and repetition specific utterances are reaccentuated and put into quotation marks.

Such utterances, it must be repeated, work in three ways: as mnemonics in establishing the continuity of a work tradition in which time is familiar, but constantly relocated in both specific places and practices. This relocation is the organisation of succession. Second, utterances work as thresholds which mark a break in the life cycle of the weaver or of the agnatic line. This threshold is the place where action occurs, a place where time appears instantaneous, partaking only of the time taken to recite the utterance. It is as if time has no duration and falls out of the normal course of biographical time. Third, such utterances are repetitive, combining either a reading or a memorized recitation. Time is specific. Repetitions recall such specificity in their unsubstitutable singularity.

In each of the four practices we discern the temporal continuity with the work tradition and the simultaneous breaks initiated between the addresser and addressee. Continuity inheres in utterances as mnemonics, while the temporality of breaks or thresholds is located in the technical (loom) and social (household) arenas. In the everyday production of cloth the work of operating the loom emerges as a competitive act posited as a right. In making cloth for the shroud this right gives way to agnatic obligation. Together, these two practices give substance to the work tradition as an act of memory and repetition. In the initiation ceremony work appears as substitution: the addresser substitutes for the addressee, best expressed in the statement, 'his time has not come'. This substitution is located in the two areas of the work shed and the mosque. This break is cemented in the transmission ceremony when the loom is gifted to the lower generation. Work is constituted as nazrana.

Common to the four practices is the recitation of supplicatory prayers found on the *Mufidul Mu'minin*. Although the use of the text may point to Bourdieu's notion of strategies since it refers to the repetition of an assumed world, it must be mentioned that the text is employed because it discursively codes a knowledge of what weaving serves and stands for. Its use by weavers refers to a written tradition. For this reason, the view that what actors do means more than what they say (Bourdieu 1977: 79) must be abandoned in favour of an approach which considers what they say to be of equal importance with what they do.

Chapter Five

Women's Work: Quilt Making and Gift Giving

Introduction

From within the world of weaving, women are constituted as workers, never weavers. This is because weaving is fortified by moral, political and legal authority and its discourse is authoritative, uttered and written. Conversely, this chapter, in focussing on quilt manufacture argues that the woman as the producer of the quilt has a status that is not formulated solely in reference to the tradition of weaving. Through the process of making the quilt the producer tells a story about herself, one which establishes a resemblance between the procedures of making the quilt and the quilt maker. That is to say, as finished product, the quilt is simultaneously a material thing and a statement about the maker of the quilt. Thus, in reading the quilt as a technical and symbolic sign this chapter argues that the quilt producer is able to assert her status independently of the world of weaving and reflect upon the work of men.

By a relationship of resemblance I do not argue that the quilt replicates the discourse of women as quilt makers. Inherent in the relationship between the quilt maker and the quilt is an intertextual dialogue with weaving which often appears to be one of conflict. The resemblance between the quilt maker and the quilt occurs through a plurality of voices, each seeking to differentiate itself within a prescribed working space. The notion of plurality denotes the character of work which cannot be framed within the limits of an established orthodoxy. However, by incorporating the quilt in her personhood, the quilt maker embeds in herself the possibility of a process by which alien, external materials may be transformed so that they reemerge as a function of her nature. This process is the technique of making quilts. For this reason the way quilts are made is of fundamental importance in understanding the relationship of the quilt maker to the quilt and through it, to the way work in the domestic domain is constituted.

Unlike the case of weaving, the practice of quilt making is not subject to a rigorous structure of conventions in which categories of people are spatially and temporally fixed. The exception to this lack of fixity is the woman head of the hearth, known primarily as one who makes quilts. Just as for weaving, the act of quilt making is divided into two broad domains: quilt making in everyday life, and on the occasion of the marriage of a woman member of the hearth. The significant contrasts between the two refer, as in the case of weaving, to the type of person who makes quilts, the uses of space, of different temporalities, of colours and designs, and of the word. More important, the act of quilt making considered in its entirety shows that work partakes of an element of play, and stands in sharp relief to the work of weaving. Collating these meanings together it is clear that work in the domestic domain is organized along the area of production and reproduction (in the sense of recycling waste material) and of gift. The gift relationship is seen on marriages. Before I discuss this I will delineate the different types of quilts, the implements used in making them and the working procedure.

The data I have for quilt making has been gathered mainly from one hearth, that of Miriam and Muhammad Umar of the Muhammadi bihaderi and perhaps reflects the bias of its occupants, especially its woman head. Wherever possible this information has been cross-checked with those of different villages. My access to the hearth of Miriam and Muhammad Umar[1] was almost total in that I was, for the period of my stay, housed in the zanana, where I was recuperating from an illness and where I was initiated into the intricacies of selecting varieties of worn-out clothes that compose the quilt. I was shown how quilt making forges kinship ties between this hearth and the kumba.

---

[1] Miriam and Muhammad Umar are two of the oldest members of the Ansari community of Mawai. By his own admission Umar is more than seventy-five years old and his wife, too, is in her seventies. They live in their own hearth with their son, his wife and children.

The Quilt

The quilt is a multi-purpose article, used for sitting on the floor or on a bed. In the cold season it is used either as a blanket or as a shawl. It can be folded four times and made into a soft mattress for a child's cradle. The more ornamental quilts, invariably embroidered, are taken out for use on festive occasions.

In general, the quilt is rectangular. Its outer coverings are made of white or green cloth stitched together. Between these layers of cloth rags are distributed evenly. The quilt comprises two layers of fabric stitched together, often with a filler in between. The quilt is made by long running stitches, except for the borders, which are folded and hemmed. On both the small outer coverings small coloured rags are arranged and appliquéd with hem stitches. Alternately, the marriage quilt, made of two or three coloured pieces of cloth, is also embroidered.

It is important to technically distinguish cloth woven by weavers and quilts stitched by women for everyday use. In principle, a woven fabric occupies a space marked by light grooves since the warp and weft intersect perpendicularly. Second, while the warp is fixed, the weft is mobile, passing above and beneath the warp. Third, this fabric is necessarily closed in on at least one side: it can be infinite in length, but not in width, since the latter is determined by the frame of the warp. Finally, the fabric must have a top and a bottom even when the warp and weft are the same in nature, number and density, for weaving reconstitutes a bottom by placing the knots on one side. In contrast, the quilt is a supple product which implies neither a separation of threads, nor an intertwining. As finished product, the quilt is an entanglement of various fibres obtained by subjecting them to moisture, heat and friction, consequently resulting in their shrinkage. This aggregate is smooth but not homogeneous and contrasts point by point with the space of woven cloth. In principle, it is open in every direction because it does not have fixed and mobile elements, but distributes a continuous variation.

In the weaving household two types of quilts are made. The first is stitched out of a variety of useless rags that were formerly clothes. The woman head of the hearth and other women members of her kumba make these quilts to be used on an everyday basis. The second type of quilt is composed of two or three pieces of fine cloth, usually selected for the quality of their texture and colour. These quilts comprise an essential part of the bride's trousseau and are made by select women of the kumba. The difference between the two is seen both in the design and colour combination that composes each of them and the uses to which each is put.

The quilt used in everyday life, a patchwork quilt, differs from that made on the occasion of a woman's marriage characterized by a central theme or motif, and often partially embroidered. The patchwork quilt displays equivalents to themes and symmetries similar to the embroidered quilt, but there is no centre. Its space is amorphous and its basic motif is composed of a single recurring element which plays on the texture of the fabrics. Visually, the patchwork quilt is nomadic. The embroidered quilt has a central motif. Along the edges of the quilt this motif is repeated on a reduced scale or reinterpreted.

Everyday Life: The Patchwork Quilt

The patchwork quilt differs from the embroidered one in three basic ways. First, the difference between the two quilts is seen in the cloth and clothes that compose the patches, the designs represented on them and the personnel who work on the two quilts respectively. Second, from the perspective of quilt makers the working procedure of the two types of quilts is different, especially in reference to the space and temporality of quilt making. Third, the embroidered quilt embodies a notion of gift not found in the patchwork quilt.

In the patchwork quilt the following materials are used. Headgear (*pagadi*) used by males on formal occasions, such as, marriage. It is of fine quality and usually cream or white in colour; sometimes, if the male possesses a turban (*dastar*), that too is used. Made of fine cotton, it is invariably white in colour. Shoulder cloth (*gamcha*), used as a scarf and towel, is usually woven by weavers and is a combination of two or more colours. A coarse cotton sheet (*chadar*) used by weavers as a shawl, is almost always white in colour and often that cloth that weavers were unable to sell in the market. Sometimes cloth for the shroud is also called chadar, but it is of a different texture and strength. Used also is an upper body garment worn like a shirt (*kurta*). Usually its chest part is utilized. While cloth for the kurta may be woven by weavers, tailoring is not done by them. Consequently, it is bought either from the market or the Gandhi Ashram. The kurta, white in colour, is often a combination of white with some other design. Finally, the upper part of the *pajama* and its leg and thigh portions are utilized. Both the kurta and pajama are made of coarse cotton cloth.

Among the clothes of women the following apparel are used. A long white or green skirt (*ghagara*) is used. This dress is often made of fine cotton and of bright colours. Occasionally, when the ghagara is made of synthetic material quilt makers complain about the distortion of the design (*naqsha*) of the quilt. Also used is the hooded cloak (burqa), usually black or white with two eyeholes. It consists of a cape veil and a coat. In the weaving community individual ownership of burqas is rare. Most often, the coat is used. Blouses, occasionally of silk and bright in colour are also used.

The above is the basic material necessary for quilt work in everyday life. To sew quilts, the rags must be first washed and beaten in an iron trough (*dabua*), shaped like a washing basin. When a woman of the hearth is getting married the iron trough is replaced by a seasoned earthen pot (*kadha'oni*). Patchwork quilt rags are washed in washing soda bought from a nearby town, while the pieces of the marriage quilt are cleaned in gray clay and sand (*du'mat*) that comes from a neighbouring village. Irrespective of the occasion the woman head washes rags.

The materials necessary for stitching the patchwork quilt include thread (*potli*), but not yarn (sut). Usually three pieces of white handspun thread are twisted together to give it strength. Machine made yarn is commonly used, but almost never for the embroidered quilt. Needles (*sui*) about one and a half inches long, bought from a nearby town and a curved iron blade (*kattari*) fixed into a wooden handle are used, the latter as scissors. A male member of the hearth operates it for the patchwork quilt. For the embroidered quilt a woman works the sickle. A wooden pestle, used in pounding spices and beating rice, is employed in rolling the quilt.

Working Procedure of the Patchwork Quilt

Quilt making has two distinct stages: washing and stitching. At the outset all worn-out clothes, tied in a bundle, are laid out on one of the floors of the zanana and carefully selected for their tensile quality. To determine the strength of the cloth Miriam would ask me to hold one end, while she tugged sharply at the other. She made three piles of rags. All the coloured ones were arranged in one heap, while those that were black and white comprised the other two piles. Having arranged the rags she made ready to wash them. The rags were of silk, cotton or wool, never new. Not all rags were from the outworn clothing of men and women. As a rag the brassiere is a special favourite. Occasionally, some rags are obtained by exchanging pieces from kumba members. Sometimes, a newly established hearth, in dire need of patchwork quilts, will buy rags from a local tailor.

The patchwork quilt is washed in the compound of the zanana during the day. Work is completed before nightfall since the weaver's section of the village does not have electricity. Unlike cloth that is being woven, the time-span in making the quilt is short and concentrated. However, since these quilts are not meant for commercial purposes and since most hearths have a ready supply of patchwork quilts, they are not put to immediate use. For this reason weavers often express the view that stitching patchwork quilts is the preserve of lazy people. In actual fact quilt manufacture is extremely arduous.

Washing and Shaping Rags

After Miriam arranged the rags she soaked them in a trough containing hot water and washing soda. The rags were washed on stone slabs placed in a small enclosure in the zanana. This room served both as a bathing area for women and one where kitchen utensils were placed. Occasionally, Miriam would use soap in removing grease. Since the rags had to be rinsed in flowing water her son was assigned the task of providing buckets of water from a well situated outside the hearth. For squeezing rags she took the help of her son's wife. After the rags were washed they were taken outside the hearth, weighed by pebbles and left to dry. The son's wife arranged the rags to dry according to the three piles Miriam had made while selecting them. When the three heaps were almost dry she brought them back into the compound. Meanwhile Miriam had fashioned the wicks and strings to tie up the quilt.

Quilt Stitching

The rags were taken to the quilt room where the stitching was done. On the floor of this room Miriam arranged the long pieces of cloth which would fashion one of the outer covers. For this she selected the strongest material and arranged it in a rectangular shape, in the form of a quilt. Miriam did not take any measurements. Meanwhile, her husband sat in one corner of the room, tearing and cutting an old quilt into pieces of various sizes. Miriam would occasionally advice him in selecting cloth and ask him not to waste rags. Umar's method of cutting was to fix the sickle between the two soles of his feet and cut the cloth carefully, akin to sawing a log of wood. The cut rags were arranged only for the outer covering. Miriam did not complete the entire covering because she felt the pieces would lose the form of a quilt.

Subsequently, Miriam sewed the selected pieces together. Her selection of pieces was determined by three counts. First, she examined the durability of cloth. She said, in her experience, the centre of the quilt wore out sooner than its sides. Second, she examined the colour of the rag. It was necessary that the centre piece be of a colour that would not fade. Miriam did not want the quilt to be composed only of white rags because, as she reasoned, it was not her task to stitch cloth for the shroud. In selecting rags for the outer coverings care was taken to arrange them according to the shade of their colour. For example, there were the near whites and the off-coloured ones which had become either yellowish or rust coloured. These pieces were arranged in one group so that there would not be a sharp break in the movement of the colours on the quilt. Finally, she examined the size of the cloth. Miriam selected the larger pieces for the centre to help in the durability and display of the quilt.

To sew the quilt Miriam squatted on the floor and began with the quilt's hem. She folded the rougher edge of the lower covering of the quilt and placed it in the upper covering. She kept the folded edge in place with one foot and placing her left hand under the quilt, held up this folded edge while hemming it with her right hand. She threaded the needle herself, but occasionally one of her granddaughters threaded it for her. She sewed horizontally (the small side of the quilt) from one end to another.

Having ordered the lower side she tacked them together. To avoid creases she placed sticks and stones on the edges and in the middle. Umar then turned over three baskets of rags on the floor. From these Miriam selected and arranged smaller rags. She found it difficult to keep the rags in place, both because they were small and her arthritic fingers, stiff with age, could not move nimbly. After completing this, Umar helped her in spreading the longest pieces of turban cloth and a gamcha over the arranged rags. On the top of this she arranged the larger pieces for the top covering. She did not sew all the rags together, but only tacked the larger ones on each other. Subsequently, she began sewing the border. With her left foot she pressed the border to keep it intact. Initially, she folded the hem of about one inch with her left hand. With her right she hem-stitched the folded border. The stitches were widely spaced. As she continued stitching she moved up along the quilt. When she reached the corner she started on the horizontal side. While hemming the border she often changed a patch on the cloth, put in rags or removed them where the quilt seemed thin or thick. When she had shaped the outline and fixed the top pieces together she selected small rags to mend the worn-out parts on the patches of the outer covering.

After satisfying herself of the quilt's shape and size she stitched bits of coloured rags for appliqué work. She turned their rough edges on the outer covering and hemmed them on. She began from the left, went upward, then to the right and finally down, fixing either a square or rhombus shaped appliqué patch. The appliqué work on the quilt contained whatever coloured rags came to hand without regard for colour symmetry, so that there would be one black square appliqué, or three or four red rags crossed by a white arabesque, a brownish appliqué and a few bright green colours. For the longer bits of rags she selected two different coloured ones and placed them crosswise on the quilt for appliqué work. She had no definite plan, but while doing appliqué she would select, change and select again.

When Miriam had completed one side of the quilt she asked

her husband to help her turn it. The quilt had not been moved from the floor. Now, husband and wife took hold of the opposite corners of the quilt and with fluidity of movement, but assuring the rags were not displaced, crossed arms bringing the other side of the quilt uppermost. Miriam commenced her appliqué on this side, again with the same designs. For the centre pieces she sat on the quilt. Having cut her thread to the length of the smaller side of the quilt she knotted one end and ran a tack along the hem on all sides. In this way each row was sewn with fresh thread. Thus, the two outer coverings and the small rags were stitched together. After stitching about a quarter of the quilt she changed her method. She took its middle in her hands and stitched an irregular rectangular spiral through it. But since the old woman could not move easily over the quilt she lifted the middle slightly and carefully so that the rags would not be displaced. Having done this Miriam could fold the quilt. It was completed, ready to be stored after being washed.

In making the patchwork quilt the woman head expresses through both verbal symbols and nonverbal gestures her statement about the nature of work. As in the case of weaving, this statement is embodied in the persona of the woman head. It may be traced through two different modes by which the patchwork quilt is fashioned: the body techniques of quilt making, discerned in the relation of the body to the work process and the approximation of measurement, colour and design and the type of conversation that occurs during quilt making.

The Body Techniques of the Patchwork Quilt

The technique of stitching patchwork quilts is not as formally structured as is the technique of making cloth. Here, a relationship of symbiosis, expressed through the metaphors of the body, is established between the woman worker and the implements and objects of work. By a relationship of symbiosis I mean that the bond between the quilt maker and the quilt is one of interdependence. This mutuality is often articulated through the idea of movement, in that the quilt accompanies women on all their important journeys. As far as the patchwork quilt is concerned it speaks of a mode of distribution in a space that is ideally without enclosures; in the case of the quilt maker this space is marked through her acts. Unlike the case of weaving, where the body of the worker and the implements of work are in harmony with each other, in quilt making the relationship of the body to the implements of work is of pain, expressed both in the act of stitching quilts and its embodiment in the woman head.

None of the implements is used exclusively for stitching. Consequently, a rigid principle authorizing the use of a particular set of instruments does not exist. Unlike weaving, the quilt maker does not arrive at an organic connection between her body and the implements. Instead, body movements are staccato and dispersed in that different parts of the body carry out unrelated tasks. The nexus between the body and the implements is of dominance: the body prevailing over the arduousness of the task and thus restoring to cloth that has died a new lease of life. Miriam used the phrase 'back breaking' (*kamar tod*) to describe quilt stitching. It sometimes entailed shedding 'blood from the fingers' (*ungal ka khoon*). The initial stitching of the quilt is often expressed in terms of painful labour (*dukh bharna*). Miriam described the fully stitched quilt as cloth brought back from the dead. The quilt maker squats on the quilt and simultaneously moves the limbs of her body in different directions. In this way, Miriam explains, quilts are fashioned since unlike cloth that is being woven, the quilt must first be given form before it can be stitched. This form is kept in place by different body parts.

As in weaving the upper part of the body is divided into a left and right, but not the portion below the waist. The right hand of the worker stitches the quilts and the left either holds up the quilts or smooths the uneven difference between rags. Further, while the left hand feels the rags for texture, the right examines them for their strength. The quality of appliqué is determined with the left hand by feeling the degree of smoothness between two rags. While doing such work Miriam would often rip apart a rag, but only after having felt the quality of stitching.

Squatting is a second typical posture of the body, which becomes a technique. While squatting all the limbs of the body are in irregular motion. For instance, while the left foot is extended towards one corner of the quilt, the right is implanted in the centre. Depending upon the part of the quilt being sewn the worker pivots the right foot and keeps the rags in place. Miriam says the squatting posture is the most difficult to learn since the stress on the back is enormous. On the other hand, this is thought to be good exercise for it loosens the bowels.

Measurement

In the course of making the quilt Miriam had a definite frame in her mind. When I asked her how she was able to measure each side of the quilt without a measuring instrument she said that all quilt makers arrived at the frame (*dhancha*) through experience (*tajarba*). Second, each woman discovered her own mode (*tarkib*) of composing the quilt. Speaking of herself she found that even when she had arranged the cloth patches in the form of a quilt the frame in her mind usually altered. Miriam's speech and her physical gestures of making the quilt pointed to an imaginary conception of the finished product, evident in that she had some idea of the frame of the quilt, but an idea that was modified in its stitching. Equally, this imagination referred back to the world of material things in that her style of making the quilt was fashioned by her experiences.

In the entire process of making the quilt Miriam did not use a single measurement with one exception, but she often ripped apart what she had just stitched, or re-arranged an appliqué patch in a different way. In terms of the initial structure of the quilt, measurement was arrived at by a mental and material seeing. This is different from measurement found in weaving where the warp beam already determines the width of cloth.

Once Miriam had started to stitch she employed the same principle she had for determining the frame. Often she would compare one rag with another to arrive at the size she desired. After sewing the rag she would examine it, sometimes rip it apart and restitch it after altering its size. The only time she employed a physical measure was when she was hemming the sides of the quilt. Here she measured the hem to a size of a digit of her index finger and then stitched the folded part to the rest of the quilt. The hem, she explains, must be stitched in this way, or else the arrangement of the rags goes awry. In effect, a part of the quilt must be fixed, for in the absence of this fixity quilt stitching can continue indefinitely. The movement of rags is infinite in length and width since there is nothing intrinsic to their arrangement that suggests a cut-off mark. The hem is important for this reason:

it provides a limit beyond which rags cannot extend. The hem is an enclosure and the only static factor in the quilt.

Colour

As in the case of measurement there was an apparent ad-hoc quality to Miriam's arrangement of coloured rags. We have, however, seen that the initial separation of rags was based on colour. Miriam was constrained by the number of coloured clothes available as rags. In the course of observing her work it became evident to me that it was not the colour of individual rags that was important, but the overall colour scheme of the quilt.

As far as the colour combination was concerned the quilt was divided into two sets of colours arranged lengthwise. One side had a white base with coloured rags appliquéd on it, while the other had a coloured base with similar designs tacked on it. Miriam explained that the side with the white base was reserved for the menfolk of the community while that with the coloured base, most often of green or pink, was used by the women of the hearth and the kumba, as well as close relatives living outside it. The emphasis of this colour distinction reflected not only upon the division of men and women, but also the public and undifferentiated world of men, on the one hand, and the private, interior and exclusive world of women, on the other. Further, as far as possible, the men's side of the quilt is composed of male and neutral garments, and occasionally of the outer clothing of women, while the women's part of the quilt is composed of women's apparel and the undergarments of both men and women. When I asked Miriam the reason for these distinctions she said that it was according to custom (*dastur muafiq*). Cloth woven by weavers acquired vigour (*jan*) after it was inhabited by people. In accordance with this quilts were stitched.

While visualizing the colour scheme of the quilt Miriam devoted maximum attention to the centre. She arranged her quilt so that the purest colours gravitated towards it. The cumulative effect was of a smooth but non-homogeneous movement. When the fully complete quilt was visible it did not have a centre, in part, explained by the diverse quality of clothes used as rags. While these rags gave the quilt a non-homogeneous shape, Miriam took pains to smooth them to a uniform thickness. When I remarked on this peculiarity Miriam said the centre was a point of reference in her head: it supplied the principle of terminating the work. Therefore, quilts are always stitched from the borders towards an imagined centre.

Design

Once the quilt was stitched to a definite shape Miriam commenced her appliqué work. Just as she had done for the filler, Miriam was particularly careful about maintaining the smoothness of the appliqué patches. In selecting them she did not emphasize the colour of the rags as much as the overall design she had in her mind. For this reason, her selection was based on the size and thickness of the patches. After isolating the rags of the same size and density she picked those patches that came to hand.

The designs used by Miriam were geometric. These were the cross, square, triangle and rhombus. They were not accorded a fixed space on the outer covering of the quilt. Instead, there was a spontaneous quality of representing designs. In all the quilts I have seen, the two sides are not identical in terms of the location of each design. Miriam said that until she had felt the rags and arranged them on the basis of their size and density she did not know which designs she would appliqué. Sometimes she would change one design in favour of another using the same rags as she had for former designs. When asked about the basis of such change Miriam said the rags spoke to her differently. Later she said these rags were associated with those who had previously worn clothes out of which rags were now made. Developing further on this theme she talked of her first effort at quilt making and the resulting fiasco due to her inability to see beyond the materiality of the rag.

The haphazard distribution of designs over the outer coverings of the quilt makes it difficult to establish its centre and periphery. Also, since one appliqué patch runs into another the cumulative effect is of a movement of colours and geometric figures. It is almost as if movement is predicated on a principle. Miriam said the anchoring of designs was found in the thumb and index finger of her right hand which held the stitching needle. She said in her natal hearth she had learned these movements, but only in her conjugal hearth, after she had become its mistress, was she able to consolidate this craft. As she put it: 'It came with my life, it will leave with my corpse (*Jan ke sath ai hai, janaze ke sath jaegi*)'.

Dialogue

The nature of dialogue in quilt making is different from that of weaving, where all verbal statements are formalized as *du'a* and are repetitions of a sacred formula found in a written text. In stitching the patchwork quilt the dialogue Miriam carried out was part of everyday speech and consequently reflected her concerns of everyday life. She began by explaining to me the details of making the patchwork quilt, the materials used, the type of cloth to be selected and the nature of such work. While selecting apparel to be used as rags she would often recount the genealogy of a particular piece in a way that the quality of cloth, as she perceived it, also reflected on the nature of the person/s who wore this cloth, as well as the type of person who could have worn it. Cloth made out of synthetic material was subject to her withering comments. In her estimation people who wore such clothes were *bazari* (of the marketplace, meaning people who sell themselves). This was in obvious reference to her eldest son who, after having come back from Bhiwandi, established his own tailor shop in the village and became relatively prosperous, but refused to help his parents in distress. He was noted for wearing brightly coloured synthetic shirts.

Holding a worn-out kurta to me Miriam explained that since its chest was more frayed than its back, it was reasonable to assume that the person lay more on his stomach than on his back. The sexual implications of such a phrase were made apparent thus: 'The kurta has some heat (*kurte men kuch hararat malum hovat*'. Hararat here was used in the sense of passion). These statements were made in jest, but often pointed, through irony, to Miriam's relationship with a particular member of her hearth. In another instance, examining a new, but ripped apart brassiere she remarked: 'The hand moves excessively (*hath bahut chalat*)'. On being asked whose hand moved thus she mentioned some distant relative of the kumba. Since brassieres are not exchanged for quilt making, not even among close consanguineal relatives, it may be assumed that Miriam was referring to a hearth member.

On another occasion Miriam told me why her selection of a

formerly white sheet to be used as rags was an appropriate one. First, it did not protect the weaver from a draught of wind since there was a gaping hole in its centre. Second, the untorn part showed it was well woven. Third, it had been used by a good man (obviously her husband), who was discerning in his selection of yarn that was to be woven into cloth. But the most important reason was that the quilt had to be balanced with rags of men's and women's garments, since her husband and most other weavers she knew refused to sit on that part of the quilt fashioned by women's garments. Thus, while stitching she composed one side of the quilt with women's garments and the other with men's. Never once did she directly address a woman member of her community.

Miriam's dialogue with absent members of the hearth began with stitching. Having transported the rags to the quilt room she asked her hearth members not to interrupt her. She would eat in the quilt room, not the compound of the zanana. Once the quilt was arranged into its basic form she requested her husband to tear the larger pieces of cloth into appropriate rags. She insisted that Umar (equivalent here to the head weaver of the hearth) preside over the final destruction of woven cloth. She thus distinguished the work of quilt making from weaving. Through out the duration of stitching she maintained a constant conversation about the difficulty of the task, her old age, the impending marriage of her son's daughter, the virtuous and negative qualities of the groom, the nature of my illness and so on. This conversation was conducted with me.

In a second type of conversation I could not participate as a direct addressee since Miriam directed her comments to hearth members. In the presence of her husband she talked of herself as a great sufferer (*dukhyari*), how she and the quilt were partners in sorrow (*dard men sharik*) and how other hearth members were oblivious to her pain. These comments were interspersed with firm advice on how to cut the rags. On his part Umar said of Miriam: The tongue will not stick to the palate (*zaban talu se nahin lagat*). Once he had finished his task Miriam asked him to continue his 'du'a-salam', meaning, 'go back to your weaving'. In his absence Miriam would either hum to herself or make stray remarks directed at hearth members. Referring to the *naïveté* of her youngest son, Imtiaz, she mentioned how easily he had been fleeced of his money, and then addressing him ironically as 'benefactor of the poor (*garib navaz*)', she said that after her death he would be reduced to a state of penury and beg for his food. The phrase used was: 'disgrace endured, a meal secured' (*sir par juti, hath men roti*). She compared Imtiaz to her eldest son who had broken all contact with the hearth. 'As he thrives he becomes fat' (*sukh badhe, mutapa chade*) was her comment. During this dialogue she continued with stitching. When she needed help she would stop this conversation and her body motions and shout for Imtiaz's wife. She did not call her by name, but as 'Parveen's mother' (Parveen *ki ammi*).

The various modes by which Miriam conducted her conversation were dialogic in Bakhtin's use of the term (1981: 276), where narration refers to a process of interpretation and feeling, not to a 'text' or an abstract sequence of events. Thus, Miriam would feel the rags for the duration of stitching and interpret them in that she associated them with types of people. This narration must include a voice, a point of view and the positioning of the addresser and addressee within the discourse. At the end of the story, one may have incomplete and unresolved styles and points of view, as expressed by dialogue, whereas the plot may be realized fully (Bakhtin 1981: 349).

In Miriam's narration, the dialogic voice occurred in three different ways. First, through the selection of rags Miriam carried out a conversation with real and imagined menfolk of her community. She attributed to the selected cloth the characteristics of the person she had in her mind and spoke in the voice of a woman addressing men. This was the dialogue of the quilt with woven cloth. In the second type of telling she established a dialogic relationship between herself as an individual and her hearth. Here, a dialogue occurred between an autobiography and a prevailing cultural tradition of work. This voice moved in two directions: she addressed the hearth at large when she saw herself as a sufferer. The emphasis was on her experiential situation. Later she addressed individual members of her hearth from the vantage point of her own position in it, seen in the comparison she established between her two sons. The last type of telling was established between Miriam and me, the participant observer stranger. Her story of making the quilt was dialogic in so far as it resisted a single definitive interpretation. In her pedagogic discourse about the quilt she categorically pointed out that younger women had their own way of stitching.

The three dialogic voices were not simply a dialogue, but also a polyphonic discourse, based on tellings or re-tellings of important events in her life. These tellings resonated against metonymic sites: the quilt room in contiguity with the work shed, the rags to be used for quilts that were formerly cloth woven by men. In other words, it was only in the quilt room at work on the quilt that Miriam could engage in the kind of tellings she did. The crucial difference from the dialogue of weaving was that the mode of communication was in the vernacular. In one sense at least, the referent of this communication was located outside the quilt room. Muhammad Umar was told to go back to his du'a-salam and it was him, as a male, who had to finally destroy cloth before it could be used for quilt stitching. To this extent it is possible to see the residue of weaving (in the sense of cloth cut up as rags) as having its sources in the quilt maker's mind.

We must also recognize that the production of the discourse of quilts is conditioned by pre-existing instances of the discourse of weaving. The quilt maker is also a re-producer. The problem, then is to differentiate and authenticate the reproduction. The latter is achieved not by adding something new, in the sense of the quilt, but by dismembering and reconstructing what has already been said and written in the *Mufidul Mu'minin*. In this perspective the use of alien instruments and materials for quilt making confirm authentic reproduction, but always elusively. That is to say, the materials and instruments with which the quilt maker works are never for the exclusive use of quilt stitching. In appropriating such materials the worker constitutes her own domain and simultaneously reiterates her status within her conjugal hearth. Also, the polyphony of the woman's voice arises from the multiplicity of materials used and the wide range of substitutable instruments, as also from the difference of language from weaving. The quilt maker and her quilt are always suspended within an original wholeness given in the discourse of weaving where she transits from one point to another. While making the quilt, however, the worker asserts herself as a fragmented voice: multiplicity induces centrifugality. Because reproduction personifies itself as the outcome of the producer it forges on itself an identity. The lack of a radically new discourse is compensated for by the theme of corporeal identity.

The Embroidered Quilt

The embroidered quilt is stitched by all the woman of the kumba. This occasion takes precedence over the work of weaving. If the bride is to marry within the kumba then women members of her future conjugal hearth do not participate in quilt manufacture. The embroidered quilt is different from that of the patchwork in three ways. First, appliqué rags are not tacked on to the embroidered quilt. Second, it is made in the compound of the zanana. Third, this quilt is thought to be imprinted with the mark of all the hearths that comprise the kumba, signified by the participation of the women of the kumba. This mark is considered a gift made by the kumba.

Before I discuss these differences I must mention that the space of the embroidered quilt, alluded to earlier, differs from that of the patchwork quilt. In contrast to the patchwork quilt as nomadic, the embroidered quilt has a centre. The working procedure, thus, differs. While both types of quilts embody a notion of movement, in the embroidered quilt this movement is not nomadic, where the mode of distributing rags is theoretically infinite, but migratory, where the movement of the embroidered motif is from one fixed point to another. The task of embroidering the quilt is parcelled out in a closed space: each person is assigned a particular spot on the quilt to do her embroidery. Unlike the distribution of rags, the embroidered quilt is fixed in its design: it encapsulates a non-homogeneous and sedentary space. In stitching both types of quilts we find the coding of memory. In the case of the embroidered quilt every occasion of making it refers simultaneously to a flow of time which moves from an old present to an actual present, and an order of time which goes from the present to the past, or to the representation of the old present. The patchwork quilt telescopes a memory in which past, present and future are ineluctably juxtaposed.

The use of cloth materials for the embroidered quilt differs from the patchwork quilt in terms of its outer coverings, which are composed of two or three pieces of very fine cloth, preferably silk. Miriam told me that most quilt makers utilized the dress they had worn on their marriages for stitching quilts for their daughter's marriage. Women of the kumba might occasionally provide these pieces of cloth.

While fillers are identical with those of the patchwork quilt, rags in the embroidered quilt are not used for other purposes. These fillers are composed of cloth and clothes such as the chadar, gamcha, kurta, pajama, burqa and brassiere. In arranging for this material those women who participate in quilt making contribute rags as gifts. Such gifts are known as gifts of royal yarn (*nakh shahwar*). The woman head of the marriage hearth keeps a note of individual contributions and must redeem them on a later appropriate date. She does not maintain a written record, but files it away in her mind. This filing is known as the red book (*lal kitab*). The book is notional and operates on the basis of a woman's recapitulation of past and remembered events.

Once various pieces of cloth and clothes have been collected the woman head of the marriage hearth and her sister, if she dwells in the same kumba, wash rags in the compound of the zanana.[2] Rags are initially placed in an earthen pot (kadha'oni) and rinsed lightly with lukewarm water. Later they are cleaned in a mixture of gray clay and sand. The implements with which the quilt is stitched is the same as for the patchwork quilt. The only difference is that instead of one needle being used each participant brings her own needle, and often her own thread. Miriam explains that women bring their own needle and thread because stitching on such occasions becomes an extremely competitive act.

The Working Procedure

Just as it is for the patchwork quilt, the working procedure of the embroidered quilt is divided into two distinct stages: rag washing and stitching. Initially, rags which compose the outer

[2] If her sister does not belong to the same kumba, then the woman head should, Miriam states, choose one of her consanguineal relatives living in the same kumba. In conversations with men I was told that the woman head usually asks for help from a member of the kumba with whom she shares a close personal bond.

coverings and fillers of the quilt are laid on the floor of the compound of the zanana. The woman head and her companion begin the process of selection and according to the rags available decide how many embroidered quilts will be stitched. Usually between three to five quilts are stitched, though I have heard weavers say rich hearths stitch up to eleven embroidered quilts. If the marriage hearth does not have adequate rags urgent summons are sent to kumba members to contribute the required material. This contribution, too, is filed away in the lal kitab.

Rag Washing and Shaping

Immediately after the morning meal women of the kumba congregate in the marriage hearth. Rags are washed about two weeks before the bride is married. The mark signaling the initiation of quilt stitching is the entry of the barber in the marriage negotiations between two hearths. This period corresponds to a time when the bride enters into an extreme state of parda. She is not allowed to do work connected with weaving or quilt making. Social interaction with all males, with the exception of the holy man (pir), is strictly prohibited. In local parlance this is known as, 'To be seated on the trunk of a tree (*manjha par baithna*)', when the bride is massaged with a white paste and put on a diet of milk and its produce. On the day marriage quilts are made, the kitchen of the dwelling is used exclusively to make sweets (*gul galla*). All other food is provided by the MB. This, too, is known as the gift of royal yarn made by the MB (*mame ka nakh shahwar*).

After women have congregated for washing rags, the woman head of the marriage hearth displays the pieces to be used for stitching the embroidered quilt. Miriam says the best two pieces are invariably hidden to protect the quilt from the evil eye (*buri nazar*) of jealous workers. She and her companion will stitch the quilt from these two pieces, which have been washed and dried in advance. Once the rags, with the exception of the hidden ones, are arranged on the floor, workers organize themselves into groups of two or three and commence to wash them. No male is allowed to participate in this activity. The youngest workers are usually saddled with the oldest and dirtiest rags. While such apportioning is not explicit an informal distribution operates on the basis of generation and age.

Quilt Stitching

Unlike the case of stitching patchwork quilts, washed rags of the embroidered quilt are not shifted to the quilt room. In most Ansari hearths the quilt room is a small enclosure, incapable of accommodating many people, as well as a place where the nuptial bed is spread. Miriam explains that embroidered quilts are not stitched in the quilt room because the nuptial union of man and woman constitutes the quilt room.

After the rags are washed and hung up to dry the woman head and her companion take out the two hidden rags and hold them up to the admiring gaze and comments of other women. Sweets are distributed after this display and the two women begin stitching. Henceforth workers sing their songs and often make lewd comments about the bride and groom. A favourite theme of the songs is grief and hardship. 'Where should I go? Where should I fly? helpless, wretched me (*Kit bhag jaun? Kith uth jaun? sun main be-bas dukhyari*)' was a song I heard frequently. A second type of song is directed at the callousness of the bride's mother-in-law: 'In every house she repeats the sufferings at the hands of her mother-in-law (*ghar ghar sas ka dukhra roti phirti hai*)'. A third type of song, directed at the husband is more traditional: 'The mango trees are in blossom everywhere/ My husband O sister still stays elsewhere (*Sab ban ambva manjari gailo/ Nandi, sainyan nahi ae'lo*)'.

Unlike the patchwork quilt, which remains immobile on the floor, pieces of the embroidered quilt move from hand to hand. Once tacked together, the pieces are arranged on the floor and the fillers selected from a basket of rags kept nearby. The compound of the zanana is a cacophony of voices: some women sing, while a second group laughs and jokes about the bride and groom and a third discusses kumba politics. Each group is fluid in its constitution, with women moving in and out with ease. This feast of words is supposed to be rivalled by the feast of things and later by a copious meal. This feast takes place under the sign of licence, in that there is no prescribed way of stitching and no constraint on laughing and indeed immodesty. Since stitching rags is the most monotonous and arduous task of quilt making, the younger women are burdened with it, while the older sew those pieces that are almost dry.

The division of work on the basis of generation and age conceals an ideal ground for learning the skill of quilt making. Unlike weaving, where skill is transmitted from father to son in a structured and institutional way, in quilt making skill is learned by copying a co-worker engaged in the same task. Beginning with how the needle is held between the thumb and forefinger, leading to the mode of holding up with the left hand the area to be stitched and ending with the use of the instrument for making an impression (*naqsh karne ka ala*) for the purposes of embroidery, learning occurs by imitating a worker thought to be skilled. The idea is not to duplicate the body motions of the skilled, as it is to adapt such skills to the learner's body rhythms. In this way the craft is internalized.

When I suggested to Miriam that learning occurs through duplication she said that to repeat the act of the skilled is to incur the risk of falsity since clothes talk differently to different workers. She said that the act moving the worker must arise from the most intimate fabric of the body, not from the skin. The skilled/learner contrast is thus presented as a dialogue, not a formal debate on style or ethics. It is an intersubjective contact between a tyro's work and an authentic model of this work provided by the professional. In other words, each individual worker learns through a process of mirror images. These images convey the joint notions of identity and difference, in that the learner recasts the style of the learned, but only after repeating the motions of the body.

Differences in the work of one worker from another are most clear when the quilt is embroidered. Here, different styles of learning are revealed. After the outer covers of the two sides are stitched the workers reorganize into the original groups of two that had washed the covers and the rags. They will commence their embroidery. The hem is first sewed in. Subsequently, the two women mark out the centre of the embroidered quilt with chalk, then the edges of the quilt where the motif of the centre is repeated. Until the centre and periphery are marked out the design to be represented is not decided.

Usually there is a discernible difference between the 'traditionalists' and 'modernists', with the former arguing for geometric designs and the customarily used colours of white and green. The latter prefer a variety of bright colours, with numerous flowers, petals and plants as motifs. At this point most women reveal their lal kitab, citing precedents for embroidering specific designs or cautioning against others. The lal kitab becomes important because someone will provide an example of how a design of flowers was represented on an embroidered quilt and the consequent break up of the marriage. Another provides a counter-example of how the same design blessed the woman for whom it was made with innumerable male children. The arguments are often a cover for a lack of skill in portraying particular designs. Traditionalists take recourse to the discourse of weaving, saying that representing iconic figures is sacrilegious, while for modernists embroidery is not the same as weaving. The phrase which summed up the attitude to men and weaving was: 'Between two priests the chicken becomes unlawful meat, though being torn to pieces in the struggle for possession (*Do mullaon men murgi haram*)'.

Miriam says this squabbling (*dant bajna*) continues until the afternoon meal is served, and is a way of complaining of hunger. Known as the feast of royal rags (*nakh shahwar ki davat*) the afternoon meal is bought by the MB of the future bride. This is a delicate moment for both the MB and the host hearth. Most disagreements and conflicts between this hearth and others are expressed by commenting on the quality of the food. 'To preserve one's honour (*Nak rakh lena*)' shows the importance placed on the meal.

Immediately after the meal the task of embroidery begins. Two workers per quilt is usual distribution. The two women who have marked their quilt begin embroidering the motif in the centre before they move on to other quilts. The composition of each group changes but not before the major design has been stitched. Once workers agree that a design with a specific colour scheme will be represented there is no radical departure from the pattern. If, for example, it is decided that only arabesques will be represented it is left to individual workers to arrange these arabesques in the desired way: in a rhombus, square or circle.

To embroider designs one of the two workers marks out its traces with wet chalk, while the other stretches out the centre of the quilt. The design on the centre is fully embroidered before the periphery is worked upon. To make the design each worker begins from the lower right side and moves horizontally to the left corner. For the centre motif the worker squats on the quilt, but does not move her limbs in different directions. The quilt is held in place by the co-worker, who uses a large stone to anchor one end, while she squats on the other. The form of the entire design is first traced before it is embroidered. For the embroidery both workers sit facing each other with their buttocks resting on the ground. The outer covering of the quilt lies on their laps so that the centre falls between them.

The more professional of the two takes the lead in embroidering the design. If her co-worker is a novice she learns as much of this skill by observing and imitating the professional. The learner does not finish the design of the professional but represents what she has learned from the professional on the quilt's periphery. The duplication of designs is considered the work of a poor craftswoman because, as Miriam says, rags do not speak to the worker. Hence, imitation necessarily entails the absence of that which, on the visual surface, it seeks to represent. Likewise, the designs being embroidered always interpose their opacity between the teacher and learner and all conceivable experiences of how these designs are seen by different workers. Finally, the design disguises its intermediary function as a mode of dialogic communication between learner and professional by posing as an event (of marriage). Hence, Miriam's emphasis on why designs can never be authentically duplicated by another. At the same time, the impossibility of escaping from visual appearances does not eradicate the impulse to stabilize designs in a referent located outside their domain. For this reason, traditionalists argue for representing geometric designs on quilts: they are the ones woven in the tradition of weaving.

If traditionalists emphasize the geometric designs and colour combination of weaving, modernists by representing new designs, provide weaving with new patterns. On the one side, weaving through the traditionalists, provides the reference of quilt making. On the other, in the case of the modernists, quilt making is the site of innovation for weaving. Thus, for instance, those who weave for the market have started to make floral designs.

In representing new designs in lasting colours, modernists, comprising younger women of the kumba, take recourse to improvisation. Its use disrupts premeditation: the moment of improvisation alters the plan of the previous design. Crucially, it is the moment when quilt making asserts its freedom to exercise intrinsic powers. This is seen in embroidering floral designs in orange and black. Though Miriam talks disparagingly of women who embroider in this way, she acknowledges that quilt making is not subject to the customary rules of weaving. For her, the difference from weaving is that in improvised embroidery the hand does not move along the path visualized by the eye. Sometimes these attempts are frivolous, but occasionally improvised patterns enter the lal kitab of a later generation.

The contingent nature of improvised embroidery constitutes an elusive present as the place and time of licence, most evident in a learner/professional relationship that distinguishes true from false imitation by emphasizing the authenticity of hidden skills. In imitating the embroidery of a skilled worker, learners use different colour combinations and a thinner or thicker needle, thus escaping the constraints of duplication. A particularly adventurous learner rubs out the traces of the design before embarking on her version of imitation: immediacy guarantees of authenticity. Immediacy appears as a way of doing, rather than of seeing. Because of this the movement of the hand is empowered with a voice in that it enters into the lal kitab. Once it enters the kitab it becomes an established way of doing. Thus, improvisation, which might appear as the antithesis of memory, is in fact dependent on its hidden activity: memory reconstitutes the treasure house that doing dissipates.

In the relationship between doing and memory speech is the visible surface. When women want to embroider relatively new designs, such as leaves of the mango tree, the hearth where it was first embroidered will be mentioned, the marital career of the woman for whom it was done examined in detail. In this sense designs are represented and recalled by their verbal images. Thus, if the use of improvisation depends on its instantaneous bid for presence as doing and verbal recall, it depends also, by an inverse movement, on the silent hidden circuits of memory.

The Embroidered Quilt and the Gift Relationship

The lal kitab is not limited to the recapitulation of designs. It resonates in the marriage quilt through the gift relationship. An

Case Study of the Gift Relationship of Muhammad Umar's Kumba

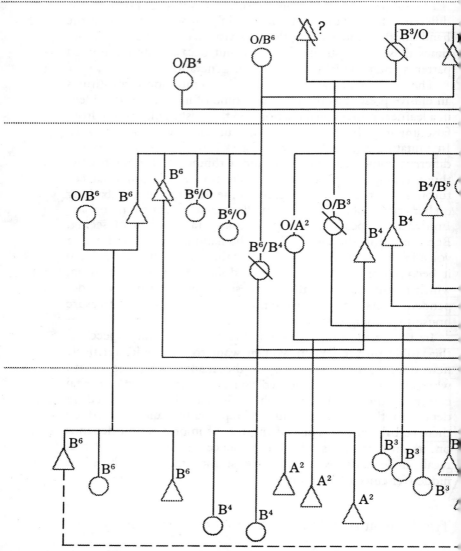

PROPOSED MARRIAGE BETWEEN B^6 AND A^1

Women's Work: Quilt Making and Gift Giving • 169

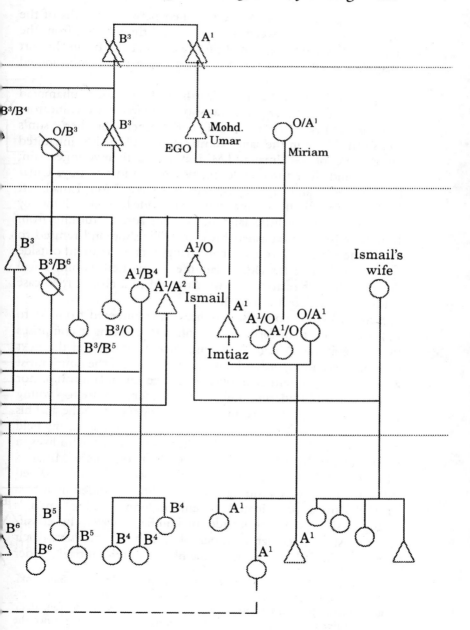

elaborate gift relationship is established between hearths of the kumba, as well as between the kumba and the bihaderi from the day invitations are dispatched till the celebrations begin. In part this relationship is structured by the act of making embroidered quilts.

My data is derived from the hearth of Miriam and Muhammad Umar. Their granddaughter was soon to marry her classificatory brother (FFFBSSS or father's father's father's brother's son's son's son) living in the same kumba. Umar's hearth is inhabited by seven people — Umar and Miriam, Imtiaz their youngest son, his wife and their three children, two girls and one boy. Umar has three sons and three daughters. His second son, Ismail, left for Bhiwandi after he was married. To provide his son with money Umar sold off the little land he owned. After returning from Bhiwandi Ismail established his own tailor shop and refused to live in the kumba. Two of Umar's daughters are married outside the kumba, while his eldest daughter is married within. I will describe the gift relationship with the aid of a chart.[3] (See case study of the gift relationship).

Umar's kumba comprises six hearths (numbered from A^1 to B^6) with more than forty people inhabiting them. It comprises four generations: Miriam and Umar's generation, and the next three in descending order. I have not represented the junior most generation. It is neither implicated in the gift relationship, nor involved in embroidering quilts, since its members are too young to work. Three people are alive in Umar's generation: he and his wife (Umar died in April 1987, a year after I completed my field work) and his FBSW (father's brother's son's wife), who lives in her husband's hearth B^4 with her youngest son. She is Miriam's FBD and was her companion in selecting rags for the embroidered quilt. In the first descending generation three women members of the kumba participated in embroidering quilts. Imtiaz's youngest sister, married outside the kumba, as well as his classificatory sister, a former member of hearth B^3, provided their services. The three women members were: Imtiaz's elder

---

[3] An Ansari wedding, in the fashion of Muslim weddings in Barabanki, is constituted in four stages: *tel*, *miyan*, *barat* and *ruksatti*, with the last stage divided into *walima* and *chauthi*. The four stages occupy four days. I will consider the four days as parenthetical to the ethnography since the quilt is stitched before the first day.

brother's wife living in hearth A^2, his classificatory brother's wife (FFBSSW), living in hearth B^5. Imtiaz traced this bond through his classificatory brother (FFB2eSS), though this member is his classificatory FBD and is also related to Imtiaz from the maternal side. This mode of tracing the relationship was determined by the occasion. The third woman was Imtiaz's eldest sister, married to a member of hearth B^4, who is in the process of establishing his own independent dwelling.

In the second descending generation female members of the kumba and one of the tailor's daughter participated in quilt making. The second daughter of hearth B^3 and the eldest daughter of hearth B^5, both unmarried, provided their services. Imtiaz traced his relationship with them through his classificatory brothers. The third female is Imtiaz's eldest sister's daughter (hearth B^4). The fact that the tailor's daughter participated in quilt making shows some contact between Ismail and his father's kumba. Imtiaz and Miriam held differing views of this contact and the value accorded to the gifts given to the marriage hearth.

Certain women of the kumba do not engage in quilt making. Two injunctions inform their participation. Unless specifically invited, widowed and divorced women are excluded from stitching. The second injunction is more regulatory than the first. If a woman is marrying within the kumba her future conjugal hearth will not share in the task of quilt manufacture. In the present instance, while Miriam's FBD, married into hearth B^4 and widowed, helped her in the task, in the descending generation of the kumba, no woman of hearth B^6, participated in the activity.

The gift relationship seen in the manufacture of embroidered quilts points to, in Mauss' recurrent theme, the inalienability of things as gifts (Mauss 1954: 9–10, 18, 24). Because of the anthropomorphic nature of the transmitted thing, 'The alliance contracted is not temporary, and the contracting parties are bound in perpetual interdependence' (1954: 62). In the above sense the gift creates a debt that must be repaid. I will take this idea of Mauss as my point of departure.

Gift exchange during quilt manufacture operates between groups and individuals. Though the latter exchange is often between two people it is not a 'personal' gift, but a restricted exchange involving two groups. The important qualification is that in quilt work the exchange of things is more important than the exchange of women. Further, the exchange of labour power, implicit in the relationship, is expressed in the idiom of the gift. By labour power I refer to the ability to do productive work in both an economic and ritual sense. I do not thereby argue that the exchange of labour is a symbolic equivalent for the exchange of women, but that the category of work partakes of an exchange where the distinction between gifts and commodities is part of the same continuum. The variable in the movement from one to another is, as Sahlins says, 'kinship distance' (1974: 185–276), which also reflects on the distinction between alienable and inalienable things.

The duration of making embroidered quilts stretches from the moment the barber enters marriage negotiations between two hearths till the first day of formal celebrations. The barber's wife carries invitations for participation in quilt stitching, while the barber carries the more general invitation to attend the marriage. The barber's wife delivers each invitation, in the form of red thread and cardamom, by entering the invitee's hearth after sunset from a back door so that she reaches the zanana without passing through the mardana. In the present instance, the invitation was carried to two hearths, B^3 and B^5, and delivered to each women head. The invitation does not ask women to participate in quilt stitching; instead it is worded as the 'feast of the rags (nakh shahwar ki davat)'. In the case of hearth B^3 the invitation was given to the eldest daughter in the second descending generation since her mother had died in childbirth. Hearth B^4 would not receive the invitation because, Miriam explained, its woman head (in Miriam's generation) had been widowed, and sending her a formal invitation was tantamount to asking her not to participate. She meant that all the concerned hearths' women must be formally asked to be present for the feast, it being understood that widowed and divorced women will refrain from coming. Invitations are not sent to hearths which share a close and personal relationship with the marriage hearth.[4] For this reason, hearth A^2

---

[4] In Miriam's estimation her personal relationship with the woman head of hearth B^4 allowed Miriam to overlook the fact that the latter was widowed. If, however, she had received the invitation it would indicate a serious breach in the relationship between the two hearths.

would not receive the invitation. Hearth B^5 received the invitation through the female barber. The invitation was also sent to the tailor's hearth despite Miriam's strong objections. The tailor's wife accepted with alacrity, and according to the barber's wife, plied her with gifts and money. An acceptance of the invitation is signaled by giving the barber's wife a quantity of flour and whatever has been cooked for the day, though she often demands money in addition to these gifts.

The next morning those who had accepted the invitation, including marriageable and married daughters, assembled in the marriage hearth. They carried with them their needles and thread and a few rags of fine quality cloth. The latter were presented collectively as the kumba's contribution for quilt making. Each woman carried a small amount of mustard oil mixed with herbs (*upton*). Meanwhile, the future bride's FZ — Imtiaz's elder sister — brought white paste and a gift of clothes for the groom, usually worn after the nuptial night. This gift is called *man'jha*, a paste also used in kite flying. After the presentation of the nakh shahwar, the rags were sorted out and washed and while they were drying the future bride was massaged with mustard oil, but only after her body hair was pared by the female barber using a traditional nail cutter (*naharni*). In return, she received a gift called *nakhur*, composed of clothes the future bride wore immediately after her hair was removed.

After the barber finished her task Imtaiz's eldest sister, married into hearth B^4, left the clothes meant for the groom with Imtiaz and went into the zanana to massage her brother's daughter. The bride was massaged with white paste in a secluded room and then bathed with mustard oil in the compound of the zanana under the scrutiny of the workers. During her bath sweets were distributed to everyone present in the hearth.

During the oil bath the MB's entourage arrived with the royal rags of the MB, which included food for the feast, water to be used to bathe the future bride, a piece of woven cloth, called *gara*, to seat workers when they began their embroidery work, and individual gifts for members of the marriage hearth. On the day of the *barat* the cloth would seat the head of the barat and his relatives. The entourage was composed of women members of the MB's hearth. They served the workers during the feast.

The entry of the tailor's wife struck a discordant note. I was told she carried with her an extraordinary hamper of sweets and money, which she presented to Imtiaz and not, as is the custom, to his wife. When he protested she gently rebuked him, saying as his elder sister she had this right. In a later conversation with me Imtiaz admitted the anguish this caused Miriam, whose displeasure with the gift was not that it was extravagant, but that it had not been presented to Imtiaz's wife.

In the area of personal gifts kinship distance expressed itself most clearly. Neither hearths A^2 nor B^4 were sent an invitation for the feast though both share a closer kinship relationship with hearth A^1, than hearths B^3 and A^5. Between hearths A^2 and B^4, however, there were significant differences in terms of the gift exchange. The woman head of A^2 made a gift of silken cloth for the outer covering of the quilt and presented the future bride with undergarments and bridal cosmetics. She also made a gift of money to Imtiaz's wife. The woman head of B^4, in terms of the genealogical position she occupied, had already made a gift (man'jha) to the hearth. In addition, she gave Imtiaz's wife some clothes for her personal use, but did not gift the bride. She scrupulously refrained from gifting money. Imtiaz expected the gift from his brother as a matter of course. This gift would be reciprocated in terms of his ability. He was not constrained to match the gift save to reciprocate. With his sister in hearth B^4, however, Imtiaz was obliged to exceed the gift, both in terms of the labour his hearth would supply to B^4, and the quantity of things given. This obligation exceeded the gift he would make as the MB. Also, in the case of hearth A^2 it was permissible to give money as part of the gift exchange, whereas if hearth B^4 had done so it would have been considered an insult.

While the gift exchange with hearths A^2 and B^4 was conditioned by the fact that Imtiaz's brother and sister are members of it, the exchange with hearths B^3 and B^5 was determined by them being classificatory relatives as well as a source of labour. Unlike hearths A^2 and B^4, B^3 and B^5 gifted identical things. These were a quantity of flour and a few metres of woven cloth. This gift, called nazr, had to be returned in identical amount on the marriage of any of their women in the second descending generation. Thus, three types of gifts were made. With hearth A^2 Imtiaz was not obliged to return the gift in equal measure. With hearth

B^4 he had to exceed the value of the gift and with hearths B^3 and B^5 he had to return the identical value of the gift.

Gifts must be physically given to the person occupying a specific genealogical position in the kinship structure. On the occasion of quilt stitching gifts are handed over to the mother of the future bride. When given to some other person, as done by the tailor's wife, a conflict is indicated and expressed. The gift exchange with the tailor's household was more complicated. Here, Imtiaz and his mother had differing viewpoints about the tailor and his family. While Imtiaz used this opportunity to reestablish contact with his elder brother's family, Miriam and apparently his wife, had strong objections. On hearing of the gift the tailor's wife had made Miriam and his wife decided to return it, but were persuaded to accept it by Imtiaz. Caught in a quandary Imtiaz did not know what to do with the gift. He eventually donated it to a fakir. At the same time he continued talking with his eldest brother, placating him by explaining the nature of family politics. Significantly, he treated this gift on par with that of hearth A^2: he would reciprocate in terms of his status.

In the entire gift relationship a written record of the quantity and value of gifts was not maintained as it is during the barat Instead, Imtiaz's wife kept a mental account of not only the quantity of things received, but also the amount of work each person did while embroidering quilts. This labour would be returned in equal measure.

The labour provided on this occasion, together with the gifts given, create a debt economy, the chief characteristic of which is not the accumulation of things, but the acquisition of obligations, to be redeemed later. While accepting the gift from his sister Imtiaz's fear was that he would find it difficult to repay, what he termed, the *ama'nat* (thing held in trust). Simultaneously, quilt work has the potential of turning the gift into a commodity. One of Miriam's refrains while accepting the tailor's gift was that embroidered quilts were not made for the market. Thus, the marriage hearth would not accept gifts given in that spirit. Miriam was expressing a common fear that gifts are frequently translated into financial terms.

The gift relationship may be seen as embodying two processes: the transformation of commodities into gifts and, conversely, the way gifts become commodities. The contrast between the two relates not only to the temporal interval involved in repaying the gift, but also to the quality of labour in stitching and its return by the marriage hearth. Cloth, gifted to the marriage hearth, has to be handwoven. Weaving hearths set aside a certain quantity from their weekly quota and make a gift of it. Crucially, this cloth must pass through the hands of women before it becomes a gift. Similarly, money is transformed from a commodity into a gift. This money, I was told, is never invested for weaving, but is used to defray expenses incurred in the marriage. Finally, apparel given to members of the marriage hearth is often bought from the market. Most such clothes eventually constitute the fillers of an embroidered quilt in some other hearth of the kumba.

Thus far I have argued that quilt stitching occupies various domains of social relationships expressed through action. These are: gift relationship articulated during stitching, the pedagogic style of transmitting the skill of stitching, the way women, as a group, and a woman in the singular express themselves while making quilts, the relationship of their body to the craft and the importance of using certain instruments and designs. The question is: how are these arenas linked? One way of showing this pattern is by exploring the resemblance between the quilt-maker and the quilt. More specifically, the quilt, by encapsulating a determinate space, codes it; the worker, by constituting this space embeds it in her personhood through an act of remembrance given in the lal kitab. Thus, quilt stitching is framed in a spatio-temporal network.

Spatially, the patchwork quilt is nomadic and the embroidered one migratory. The patchwork quilt, made with rags of various colours, sizes and densities, articulates a movement where every fixed point is a relay and can only exist as such. In similar fashion, the quilt-maker, as an individual, exists as an intermezzo between two points. She is suspended within an original wholeness given to her either through her natal hearth, or her conjugal one. She forever transits between the two. Hence, she can never make that which she inhabits her own.

The transition between two hearths, natal and conjugal, expresses the space of the woman worker. This space can be seen in the nature of her work as quilt-maker. The patchwork quilt is not the same as the embroidered one, which plays on the theme of a centre and periphery in the representations of designs. The embroidered design always moves from one point to another even if the latter is not clearly localized. Likewise, quilt workers as a collective, form a common aggregate where each worker is assigned a share in the embroidery and the communication between shares is regulated. In its clearest form this communication is evident in the pedagogic transmission of the skill of embroidery, and in the gifts each hearth carries. The worker must make her design on a closed and therefore localized part of the quilt. Thus, it may be said that a group of migrants, coming from different hearths to the marriage hearth, work together to fashion/quilts that will accompany the bride on her journey to her bridal home.

This work refers to two temporal movements. First, in deciding the type of designs to be represented workers hark back to a past to argue for or against the design. This recall establishes the continuity of time: the past must inform, in its affirmation or negation, the present mode of work. Thus we see women debating the type of design to be represented. While working on the design the continuity of time is ordered by moving from the present to the representation of the past. The design, then, becomes a technique of remembering the past. Conversely, in embroidering quilts, memory has a spatializing function in that it speaks through the representation of designs. Improvisation in this work occupies an ambiguous position, seen in the freedom to exercise the intrinsic powers of one's imagination. In emphasizing immediacy, improvisation appears to deny the past but we see that even here it is dependent on memory as a referential system. For if improvisation operates through a random combination of the elements of a design, it is evident that this randomness can only occur within a prescribed set. In other words, the recall and repetition of the form of a design establishes improvisation. Improvisation occupies an intermediate position between patchwork and embroidery. Patchwork constitutes an eternal present. Because of the composition of rags, their colour density and strength, every instance of making it is a present that can never imitate the previous patchwork quilt.

Chapter Six

Circumcision, Body and Community

While the previous chapters have focused on the working life of the Ansaris, this chapter explores the significance of circumcision among them. The analysis is divided into two parts. The first details the ritual of circumcision and the second the everyday discourse surrounding the fact of being circumcised. In the process of explaining the two sections the chapter marks out the discursive terrain of two terms, khatna and musalmani, used respectively to describe the ritual and the everyday discourse. Common to both these sections is the concern to understand how particular Muslim groups claim membership to Islam.

In the context of the ritual we find that it is not read from within the domain of Islam. When recognized, as among various tribal societies of tropical Africa (I.M. Lewis 1966; Trimingham 1964) it is explained as a puberty ritual. In other cases it is not seen as a rite of initiation at all (Watt 1965). The second position need not concern us here. In terms of circumcision being a ritual of entry, Bourdieu (1977: 225 n. 56), for example, says that the ritual is a purificatory cut protecting the male from the dangers of sexual union. Simultaneously, as part of a structure of practices, the ritual shows how individuals are socialized into the group. This chapter follows, in part, this line of argument, but also suggests that through the ritual and the everyday discourse a claim is being made to an Islamic heritage.

At the heart of the analysis of this chapter, then, is a larger question: what is the relationship of local Muslim groups, such as the Ansaris, to Islam? There could be at least two ways of answering this question. Given that the Ansaris are low caste Muslims, generally illiterate in sacred affairs, we could argue that they typify a 'folk' imagination of Islam. This imagination is necessarily limited to a local context. Consequently, the meanings generated in reference to it are also enclosed within the world of the Ansaris. Alternately, in their mode of worship we could isolate general characteristics, such as the repetition of sacred formulae, the liturgy and so on, by which they share the same properties as other Muslim groups. In this case, such characteristics would point to an orthodox imagination of Islam and would not concern themselves with the cultural location of the Ansaris. An orthodox point of view would assume that for Muslims Islam is the universal form of human experience, incontestable and invariable. On the other hand, a folk theology of Islam would show, at least partially, how people live their everyday lives and the relationship of their lives to the sacred.

El-Zein (1977: 227–54) tries to build an anthropology of Islam by negotiating the two positions. He argues that in the distinction between 'folk and elite Islam, anthropology studies the former, yet its principles of analysis resemble the latter' (1977: 246). But, he says, the orthodox and folk theologies of Islam are complementary since each defines and occasions the latter. If orthodox Islam relies on sacred formulae and an inviolable tradition and finds truth first in the Quran, the folk theologies of Islam locate meaning in nature and then place the Quran within that general order (1977: 48). In other words, an orthodox position shows that the Quran produces a context, while a folk perspective embeds the Quran within a natural order.

In either case interpretation assumes the universality of Islam. This position, el-Zein argues, is shared by an anthropology of Islam. As a counterpoint if one assumes that Islam is not characterized by fixed meanings it then becomes possible to collapse the distinction between the folk and the orthodox. This is because Islam is found in the content of what is being studied. This content articulates 'structural relations' in a way that Islam is a product of such relations. Given the diversity of cultural content it becomes impossible to speak, at least anthropologically, of a universal meaning of Islam. Indeed the 'logic of Islam' (1977: 252) is immanent in the content of the phenomena being studied. Anthropologically, then, Islam does not exist as a fixed and autonomous form.

I take the above position as a point of departure in the present chapter. Focusing on the content of the ritual as well as of everyday conversations I argue that the novice's introduction to

Islam cannot be separated from his structural position in his domestic group and the Ansari social structure.

In this chapter I do not treat circumcision as part of puberty rituals or situate it centrally within a rites of passage frame work if only because the age at which boys are circumcised in this study varies from between two to six years. I will show that the ritual, in constituting the body of the male, allows it to simultaneously enter into the life cycle of the domestic group and the community of Islam. This simultaneity is evident when we focus on the gestural and graphic inscription of the ritual on the body.

There are three modes in which the gestural and graphic unfold. I have tried to understand this operation through the term biunity. Biunity is the signature, par excellence, of the ritual. This signature confers an identity on the novice, one that cannot exist elsewhere. This identity, inscribed on the body of the novice, shows how the characteristics of male and female are combined. The medium is primarily gestural. Second, this signature refers to the way the body as an object or thing is posited. This positing, I will show, is found in the regimen of three types of signs impressed on the novice. The medium is both verbal and gestural. Finally, what is implied in the signature is the agency of the other. The other is not only the mark of the domestic group, but also the word of God writ large on the novice's body. The medium is primarily verbal.

The discourse on circumcision, however, is not limited by its ritual context, but is part of the everyday vocabulary of the menfolk of the community. This vocabulary does not refer to the body of the circumcised but attempts to verbally constitute the boundaries of 'being Muslim'. This constitution is achieved by separating Muslims from non-Muslims. In locating circumcision within its ritual matrix and as it is constituted verbally in everyday life this chapter suggests that the relation between the everyday and the extraordinary can be mapped along a continuum. Through the agency of the wound the ritual inscribes the divisions of the social structure on the novice's body, while in the everyday conversations the discourse on the wound is a mode by which the boundaries of the Muslim community are established. In the process the focal point of everyday conversations is no longer the body of the person undergoing circumcision, but the body of the male Ansari community taken as a

Circumcision, Body and Community • 181

unity. This chapter attempts to read the ritual and show the transitions and connections from and with it in the everyday life of the community.

THE RITE TO BE MALE

In the context of my field work, references to circumcision, initially elliptical, later humourous and ironic, were neither directly solicited nor framed within an ethnographic setting. I was witness to one circumcision ceremony, that too inadvertantly. Sometime in August 1985 I was staying in the house of Sadiq Ali when he decided to have his young son circumcised.[1] My presence in the house led to an ambiguous situation. Sadiq Ali was not sure whether I, being a non-Muslim was allowed to remain in the house during the ceremony. On the other hand he did not want to offend me and in the process appear mean-spirited by asking me to leave. The issue was resolved after I offered to leave his house for the duration of the ceremony. After consulting with the elders he asked me to stay. I started taking notes on circumcision a month later after a Hindu *patwari* informed me that because of my association with Muslims, commensal and residential, I was an 'uncircumcised mullah'. I will reproduce the conversation here because I think it is instructive. It points both to the body as a referential object and the verbal discourse surrounding circumcision.

Most evenings a few of us would sit around a local tea stall talking of the day's activities. Towards the end of September 1985 there was a marked difference in the content of our conversations. Some of my friends were greatly agitated over the emerging controversy in Ayodhya, barely thirty kilometres east of the villages of this field work. The conversations reflected the fear of

---

[1] The male head of the dwelling in consultation with the elders of his agnatic line decides on the date of his child's circumcision. Sadiq Ali says children are not circumcised during the 'dark' months of Muharram, Sabrat and Roza since this is a period of solemnity. During Muharram, he adds, some Ansaris are in mourning, in Sabrat old members of the community pray for deliverance, while during Roza the entire community fasts. Circumcision is usually done in the months of Id, the beginning of the new year, or in Chahullam, the most important Ansari festival.

violence touching the area. During one of these sessions I was introduced to the patwari.

P: Your name? (*Apka nam?*).
D.M.: Deepak.
P: Deepak what? (Deepak *kya?*).
D.M.: Deepak Mehta.
P: Are you a Srivastava? (Srivastava *ho?*).
D.M.: No, I am a Punjabi Khatri (*Nahin, men* Punjabi Khatri *hun*).
P: That's the same thing. Where do you stay? (*Ek hi bat. Kahan rahat ho?*).
D.M.: Sometime in Mawai, sometime in Gulharia (*Kabhi* Mawai, *Kabhi* Gulharia).
P: Oh! Do you stay in a Pasi household? (*Achcha*! Pasi *ke ghar men rahat ho?*).
D.M.: No. I stay with a Khan saheb's family (*Nahin.* Khan Saheb *ke sath*).
P: Is that so? Then you must be eating their food? (*Samjha! To unka khana bhi khate ho?*).
D.M.: Yes (*Ji*).
P: This is the first time I have come across an uncircumcised mullah (*Pehli bar humko mullah mila jisne musalmani nahin karvai ho*).

The patwari met me a few days later. I was advised not to talk to him.

P: Good day wise man. Have you been circumcised? (*Salam alai kum mian. Musalmani karva li?*). [On receiving no response] What happened? You can't read the namaz without being circumcised? (*Kya hua? Musalmani ke begair namaz nahin pad sakat?*).

I did not know that musalmani referred to circumcision. When I asked my friends the meaning of the term I was told that was how one became Muslim. Intrigued and rather naively I asked how indeed did one become Muslim? Exasperatedly, one of my friends visually mimed the operation of circumcision. Musalmani is the most oft-used term denoting circumcision. Other terms, both Persian and Arabic, are restricted to describe the ritual. The classical term *khitan* is not used, but *khatna* (to cut) is. Khatna is employed in the company of *tuhr* (to clean). In turn, tuhr is associated with other terms, most notably ghusl (to bathe), *hajamat* (hair cut) and *istibra* (the removal of the last drop of urine). In this sense, circumcision is a way of experiencing one's body, of apprehending it, of assuming it positively and fully. It is linked to the paring of nails, the removal of hair and the invigoration of the body through oil massages.

The two terms (musalmani and khatna) denoting circumcision, each of them constituting different terrains of action, are used on specific occasions. Musalmani is used to establish in discursive terms the quality of being Muslim, pointing in this specific case to the difference between Hindus and Muslims. The second section of the chapter develops this discourse. Khatna is used to describe the ritual. Before I discuss the ritual I will introduce the main argument of the first half of the chapter.

The ritual is characterized by two types of acts: the gestural and the graphic. In describing how the ritual is enacted the chapter focuses on the inscription of a ritual mark on the body. Here, what is marked on the body is already part of the body's production since it is included in the body. The body in this sense is referential: by means of a gesture an object is carved out. Hence the body is both signifier and signified. Together with the gesture, a second movement is discernible in the ritual. This refers to the blowing of the word in the initiate's right ear. Here what is marked on the body emerges from an environment external to it. If through gestures the body acquires an authorized mode of behaviour, with the word it is attached to the community of Islam. Both the word and the gesture are co-present on the surface of the body. Implicit in this ritual, then, is a triadic classification of the male body into a depth, a surface and a celestial height.

The Setting

The Ansaris circumcise their male offspring in the zanana of their household. The courtyard of the zanana and the quilt room are especially prepared for the ceremony. The western wall of the courtyard is cleared of impedimenta and made to resemble, as far as possible, the blank western wall of the mosque. All ritual connected with circumcision is directed towards the wall. In contrast to this wall, the quilt room is decorated with the choicest marriage quilts, the nuptial bed is made with new sheets and a little flour is sprinkled on the bed. An earthen pot, placed in one of the corners of the room, will be broken after the operation.

The day of circumcision is often referred to as 'barat', the day when marriage is consummated. It may be argued that the two ceremonies, of marriage and circumcision, are similarly structured. The practice of sprinkling the nuptial bed with flour, of the ritual bath under the aegis of the mother and the barber, have much in common in both cases. It is as if circumcision were a mimicry of marriage and the removal of the foreskin an anticipation of that of the hymen, or that circumcision is a preparation for deflowering (Bouhdiba 1985: 182).

Whether circumcision mimics marriage or rivals the rupture of the hymen are not issues I will examine. Instead, I will show that the ritual enables men to enter into the world of women. Far from separating the sexes, the ritual premises a unity between them. The site where this unity is enacted is the body of the novice.

The Dramatis Personae of the Ritual

Broadly, four types of people are involved in the ceremony: the novice, his mother, the female barber and those who witness the ceremony. The last includes the father of the novice, his mother's brother and all his father's agnates who have been circumcised. This includes all the circumcised members of the novice's generation. The mother and the female barber work on the body of the novice while the witnesses authenticate this work by providing verbal legitimacy to the act and are the only ones who talk of circumcision in the public domain.[2]

The Mother

A day before his circumcision the novice child is placed almost exclusively under the care of his mother.[3] She ensures his hair is

---

[2] None of the Ansari men I talked to remembered his own ceremony, but was eloquent in describing someone else's. In this sense descriptions concerning circumcision never refer to the speaker. They show how a collective memory is, through the ritual, inscribed on the body of its believers. I will return to the significance of this point in the second half of the chapter.

[3] Among Ansaris this day is marked by its relative lack of ostentation and

cut, his nails are pared and he is massaged with mustard oil. Of these, the paring of nails is the most significant. It is believed that the novice is susceptible to involuntary corruption just before he is circumcised. Dirty fingernails aid in this corruption because the devil takes up his residence in the dirt lying between the nail and the flesh. Ideally, nails are pared on a Thursday or Friday preceding circumcision as this assures wealth to the circumcised. The disposed off parings must be buried so that sorcerers are unable to play with them. I was given a second explanation for the burial: they are part of the human body and have to be buried like the body itself. The burial is performed in secrecy by the mother of the novice. After the oil massage and the paring of nails the novice is bathed in flowing water by his mother. Subsequently she introduces the child to her guests.

In the present instance, after Shabnam (Sadiq Ali's wife) had bathed her child he was made to eat cooling foods. Immediately after the mid-day meal Sadiq Ali's agnates, his own elder brother and his wife's brother and all the other circumcised male members of Sadiq Ali's kumba gathered in Sadiq Ali's dwelling. They were accommodated in the courtyard of the zanana. In the presence of the above members Shabnam introduced her son. Pointing to him she spoke as if he were a stranger. She used the formal third person honorofic *ap* and not yeh: *'Ap hain* Shujat Ahmad. *Apki umar teen sal hai. Kal ap* Hindu *se* Musalman *honge'* (Here is Shujat Ahmad. He is three years old. Tommorrow he will become Musalman from Hindu). The people gathered responded *'Bis-millah-al-Rahim'*. After this simple introduction the boy was asked to come sit at the head of the cot on which his father sat.

The most important part of the day preceding the ritual is the preparation of the novice. Some Ansaris say the novice is given his first bath this day and that this is his ghusl. Henceforth, the initiate ideally repeats this process of ablution every Friday: oil

---

display. Wensinck reports that in Mecca on the day preceding circumcision the boy, clad in costly garments, is paraded through the streets on horseback and is accompanied by footmen and his father's elderly black handmaid. The second part of the procession is composed of the boy's poor comrades. The entire procession traverses the main streets. Similarly in Egypt the boy is paraded through the streets. Dressed as a girl, he has his face covered by a kerchief. As in Mecca he is preceded by musicians. In contrast, circumcision among the Ansaris of Barabanki is confined exclusively to the household.

massage, paring of nails and bath in flowing water. The removal of pollution is not so much an attempt to excise the person of sin as it is to practice an 'ethics of the sphincters' (Ferenczi, quoted in Bouhdiba 1985: 48). Such purificatory techniques are then extended to cover a wide range of functions: eating, drinking, defecation, sexual intercourse and so on. The purified body will be discussed later. What is important is that through the first ghusl the novice's body is inscribed with a subjectivity through the agency of the mother. Every subsequent ghusl enacts this inscription.

On the day of the ritual, the mother after bathing her child, dresses him in the head-dress of a groom and leads him to the compound of the zanana. The novice is handed over to his MB. In direct contrast to the preceding day, the MB introduces the guests to the child by their genealogically appropriate term. After the introduction, the child is handed over either to his father or the eldest male of the agnatic line. This giving away is termed *nyota*. This man, in turn, places him in the custody of the female barber. Meanwhile, the mother places a red piece of cloth near the western wall and holds up a green cloth to the gaze of the assembled. For the duration of the operation the novice is seated on the red cloth. The green cloth will be used to wrap mother and child immediately after the operation.

After the prepuce is removed the child is held up by his mother so that blood from the wound runs down her chest. In this way it is believed blood and milk co-mingle in producing a healthy male. Blood and milk are considered vital ingredients of the body and it is important a proper balance be struck between the two. Balance is achieved only after the mother relinquishes her bond as it has existed with her child. This interpretation is suggested from my conversations with two women, Shabnam and Miriam. Shabnam, in her mid-thirties in August 1985, had participated in her first son's ceremony. Miriam, in her early seventies, and the wife of a Muhammad Umar, is considered by members of her community to be an authority on traditional matters.

While describing her son's ceremony, Shabnam pointed to the alienation of her son from herself. She began by saying, 'When I was small my father always said to me,'don't sit next to your classificatory brothers otherwise you will become like them. Play with other girls. In this lies your honour.' When Shujat [referring to her son by his formal name] became Muslim, I had to forget he was my *munna*, because this is what my father meant. I must pass on this *amanat* (thing held in trust. In this case her son) to his father. Soon enough I will not be able to play with him, to hold him, to caress him for he will have become male (*mard*).' I asked, 'What does becoming a male mean?' I was given, what appears to me, a formal answer. 'It means having enough blood to produce an offspring, observing namaz, fasting, pilgrimage to Mecca and alms-giving.'

Shabnam's reiteration of the pillars of Islam was balanced by Miriam's comments. Circumcision, she said, is the recognition of the co-presence of male and female in everyone. This recognition is implied in the combination of blood and milk. Every human is composed of these two elements, blood inherited from the father and milk from the mother. In holding up the circumcised boy to her breast the mother gives to her child the gift of milk, a gift that balances the blood of the father. Blood, she says, is red because it implies fire under which everything is either incorporated or ravaged. It stands alone. She compared the singularity of blood to the letter *alif*. Both imply a movement of things and people towards the sky.

However, for Miriam, blood emerges from the earth and is nurturing. The prime example of nurturance is the milk of the mother. This milk balances and often counters the excessive strength and anger given in blood. By a balance she means that if there is an excess of blood in the person s/he is disposed towards anger. If milk is preponderant the person is inclined towards corpulence (*mutapa*) and sadness or hardship (*dukh*). I asked her whether this balance operated in men and women in the same way. In her opinion, women are born with blood and milk, whereas men are gifted the latter. This gift enables them to enter into the world of women. Hence, circumcision is for those who will become men. Furthermore, circumcision is the necessary prelude to marriage for in its absence the product of the union between man and woman is either sterile or consumed by violent passions. In any case, the gift of milk is the last gift of the mother to her child. From the point of view of the child the ceremony, carried out in blood and pain, is rivalled by the nostalgia the mother feels when she contemplates her relationship with her mature son: 'khatna *ke sath apne walid ka beta ban jata hai*' (with circumcision he becomes his father's son).

This double movement, of affirmation and denial, is the paradigmatic expression of the ritual and may be its organizing principle. On the one hand, the mother's preparation of the child maintains the security of the enactment and the object of this enactment (the novice's body). Each enactment produces in its subjects the conditions of its possibility: the past appears through the act of production. On the other hand, the reappearance of this past is premised on the mother almost wilfully forgetting she shared a relationship with her child before he was circumcised.

The Barber

As a ritual specialist the barber transgresses various boundaries, most importantly those codifying sexuality. For the duration of the present ceremony she maintained a constant monologue on the state of sexuality among the Ansaris of Mawai. In her opinion, they did not know how to be men because they were inadequately circumcised. It was, she said, left to Sadiq Ali's son to rectify this woeful imbalance. Often she would single out one or other male member present and talk of his ceremony, his sexual prowess or lack of it and so on. Her monologue stopped as soon as the wailing child was handed over to her. She proceeded to examine his penis and comment on its power to impregnate all of womankind. To the delight of the witnesses she mentioned that even she, an old crone and thoroughly experienced in matters sexual, looked with envy at the potential sexual prowess of the boy. Then holding the prepuce she snipped off the outer end in one smooth motion. The child had lapsed into whimpers. During the operation not a drop of blood was allowed to fall on the ground. Subsequently, Sadiq Ali's FB whispered the azan and the child's name in his (the novice's) right ear. Immediately after, the menfolk offered prayers facing the western wall.

The barber's transgression of social codes during the ceremony succeeds because through her dramatic banter she points towards another code different from male Ansari notions of sexuality. The latter not only prohibit women from discussing sexual matters but also believe women are incapable of exercising judgement in sexual affairs. This code is sustained by a radical separation of the sexes, indeed this very separation is its prerequisite. In making her statement about the boy's sexuality the barber refers to a domain of feelings which are not merely corporeal, aspiring to pleasure and so on, but also point towards the investiture of a masculinity on the child. She categorically told me the penis was the boy and without it he was nothing.

I talked to the barber more than a month after the ceremony. I began by speaking of the ritual through circumlocutions. I recounted my conversation with the patwari, of the term musalmani and of her role during the ritual. Referring to the patwari she abused him for giving a bad name to a sacred act and said this was the viewpoint of the castrated (*khasi*), not the circumcised. Distinguishing between the two she said the former was achieved by the removal of one or both testicles, whereas khatna ensured the sympathy between the male and the female. She designated this sympathy by the term *hamdami* (literally, of one breath), for only this sharing of breath makes possible the male vision (*shuhud*) of the female and the female vision of the male. The conjunction between the male and the female has two aspects, in the male it is the shauq (passion or desire) for the female, and in the female, the realization of this shauq. Shauq, the barber adds, does not refer to two heterogeneous beings, but one person, either male or female encountering him/herself as the other, at once a biunity, something that people tend to forget. In other words, in this interdependence each obtains his/her recognition from the other.

Biunity in the sense elaborated above, is a given but one that must be uncovered by circumcision. Her role, as she understands it, is to reproduce as truthfully as possible this framed world. The operation, she says, recognizes the biunity between male and female. The operating instrument, a small and sharp blade called naharni, is used. The same blade is also employed to remove the hair of the bride a few days before her marriage. The barber says the naharni is used not coincidentally to remove excess bodily material on both the bride and the novice male child because both will experience a second birth. The naharni then is the instrument of hamdami. A sure sign of this blowing together is seen in the operation. If the act is swift and smooth the circumcised will be potent. If, however, the prepuce cannot be removed in one flowing motion he will find it difficult to marry and raise children. Further, if circumcision cannot be performed on the prescribed day the child is in danger of becoming a saint (pir). The last point was left unexplained.

In the above account, as distinct from that of the mother, the barber is situated outside the act. Her account is based on a framed portion of a prior world which she undertakes to represent accurately. For the mother there is nothing to retrieve, there is only the act by which her child becomes alien to her. Simultaneously she acquires agency over her child in that she is responsible for originating the entire system of signs given in the act of circumcision. The barber, on the other hand, is engaged in the theme of portraiture. Her portrait is organized around the removal of the foreskin, an act of violence by which the body is precipitated into an alterity. This alterity plots the progression of the boy's career in the domestic group: the labouring body, the impregnating body, the authorizing body and so on. Also, through mutilation the biographical time of the body is encoded so that to enter into the life of the domestic group is to enter into the community of Islam. That is to say, through such violence the body becomes a metonym of the social space of the domestic group and is simultaneously constituted as an imaginary space for the reception of Islam. Both the barber and the mother seek to constitute the body of the novice — the barber by recourse to a prior world and the mother by establishing agency over her child.

The Novice

From the account described so far the child acquires a gendered subjectivity only after the ritual. I will describe the inscription of this subjectivity by focusing on the way his body is constituted.

A day before his operation the novice is the centre of attraction in his father's house. His mother massages him with mustard oil and he is segregated much like a bride just before she is married. The phrase describing this condition is manjhe par baithna (to be seated on the trunk of a tree). The oil massage is followed by the paring of nails and immediately after by a vigorous bath. After the bath he is formally introduced as a Hindu to his father's agnates and his MB. On the day of the ceremony the novice undergoes the same regime of cleaning. Subsequently, he is handed over by his MB to his father, who then places him in the custody of the barber.

Circumcision occurs once he is handed to the barber who comments on his sexual potency. The operation is accompanied with cries of Bismillah-al-Rahim from the witnesses. The mother then holds up the bleeding boy so that his blood can run down her chest. The blood is not allowed to fall on the ground. The mother and child are draped over by a green cloth. The entire ceremony is executed in reference to the direction of Mecca. The excised part of the prepuce, surreptitiously recovered from the barber after haggling, is buried under the nuptial bed. Meanwhile one of the witnesses, called the khidmatgar whispers the azan in the boy's right ear and then recites his formal name. After the recitation the menfolk offer prayers facing the western wall.

There could be multiple interpretations of the details through which the novice's body is marked. We could, for example, show the profound relationship of the ritual to marriage, or from a different perspective, focus on the problematic reconstruction of 'facts', in that this chapter is dependent almost solely on representative testimony. The interpretation I follow is suggested in the accounts of both Miriam and the barber. I will argue that the boy's body is fashioned as a biunity. As I understand it biunity has three dimensions. First, it refers to the conferral of a gender on the child, one found in the combination of blood and milk and male and female. This gendered identity is available in the accounts of the mother and the barber. Second, biunity establishes a relationship of substitution between the physical object of the body and the regime of signs that play on it.[4] Finally,

---

[4] These signs indicate, manifest and signify. As indexicals they function by associating particular words with particular objects or images. In themselves these terms are empty of content, but are filled in through a relationship of externality and referentiality. That is to say, there is a strict relationship of referentiality between istibra and the penis, between ghusl and the purified body and indeed between the name and the person. Further, the terms are indexical in a special sense: they form material singularities by indicating how each male body is to conduct itself. However, ghusl, istibra and the name are more than indicators. Through them the novice reflects a socially recognized masculinity as well as constitutes for himself the domain of the personal. The third dimension of this regime of signs, signification, relates each of the terms to general or universal concepts. Here, signification

biunity establishes a relationship of identity between the spiritual and the corporeal.

In terms of the three dimensions the novice's body is simultaneously a system of signs and the object which is marked, worked upon and quite literally produced through the ritual. The regime of signs can be understood through three terms: ghusl (ritual bath), istibra (cleaning the last drop of urine, preferably on a stone) and kalimah (the word, to be understood as one through which the novice is impregnated with Islam). Together these three terms constitute the state of being pure (tuhr) and in this way point to the way a masculinity is conceived. As an object the body provides the depth on which the state of being pure is enacted. It is almost as if the body has no surface, no inside or outside, no container or contained. In other words it does not have a precise limit. Furthermore, as an object the body is fragmented and dissociated. In mapping the three terms onto the physical object the body acquires visible social organs and an authorized code of conduct. To develop this argument I will focus on the three terms.

Ghusl

As the major purificatory ritual of the body the importance of ghusl cannot be overstressed. Miriam was clear on what ghusl attempts to remove. Its main object is to eliminate dirt. The prime source of dirt is the human body, a dirt that is dangerous since it impinges on the tidy insularity of the body. Dirt is composed of secretion and any contact with it is contaminating. For this reason a special regime of hygiene is associated with the secreting areas of the body: the armpits and the genital zones in particular. In adults the hair from these areas is assiduously removed. The second fear, after contamination, is of decay — a rotting in the depths of the body, which to be excised must be brought to the surface. All bodily refuse is a sign of decay — urine and faeces for all humans. In addition, for men, nails that are pared and hair that is cut is also a sign of decay. In men indications of decay manifest when they are unable to procreate, while in women

---

implies promises and commitments to both the domestic group and the community of Islam.

decay reveals itself in the ability to menstruate regularly (pregnant women are of course exempt).

The ritual of circumcision is the first attempt to rid the body of dirt and decay. The paring of nails, the first hair cut, the oil massage and the vigorous bath that follows are to be understood in this light. The massage brings to the surface the decay in the depth of the body, while during the bath particular attention is focused on cleaning the anus and the genitals. Henceforth, the male child is taught to remove all traces of excrement by means of water.

Istibra

One of the many meanings circumcision lends itself to is the removal of impurity associated with urine. The removal of the prepuce prevents the residue of urine and sperm from accumulating inside the body. Immediately after the operation it is thought to be a good omen if the boy urinates (which I am told happens almost invariably) for it indicates he is in good health. On all subsequent occasions the boy rids himself of the last drop of urine by an elaborate technical procedure. It is inadvisable to urinate standing. One must urinate with the buttocks resting on the ankles. After urination the penis, taken by the left hand, is rubbed several times against a stone. Istibra is continued until nothing remains in the urinary tract. The removal of the prepuce, however, is not merely an attempt to remove the body of impurities but also the first step towards conjugal union. This is suggested since istibra is especially recommended for the recently married male. Here istibra is thought to be an important way by which the male cleans himself after sexual contact.

Lawful conjugal association between man and woman presupposes the circumcised male. If we focus on the bride during her nuptial night it is possible to infer that her deflowering, carried out almost publicly in the zanana, is equivalent to circumcision. The bride, escorted by the groom's father's sister, is led to the nuptial bed and formally introduced to the groom. She is made to sit on the bed while the woman of the zanana throw flour on her and sing songs. The groom is asked to come sit by his bride while the women congregate outside the nuptial room. Sometime during the night the nuptial sheet is put up for display, showing that the bride was a virgin, or else she is forever branded with a 'redhibitory defect'. In this sense, both circumcision and the deflowering of the virgin are marked by a cruel wound, a kind of forced narcissistic experience of oneself (Bouhdiba 1985: 187).

Here circumcision is an initiation into legitimate sexual desire, but one fraught with negative consequences. This is evident in that all the areas of the body producing secretion are imbued with a negative attitude: any secretion from them is a sign that life is anxiety and danger. The ritual in effect teaches how the danger can be resolved. Bouhdiba says that circumcision is ' . . . a vaccination against the dangers of sexuality' (1985: 185). Circumcision and the deflowering of the virgin, then, occur within a frame in which festivities, blood, pain and exhibitionism accompany the traumata wittingly inflicted by the group to maintain its own cohesion. The problem with this analysis is that it posits a perfect symmetry between deflowering and circumcision. Such a relationship becomes possible after Bouhdiba excises religious legitimacy from the ritual of circumcision and instead considers it to the extent that it reflects on sexuality. This is not to deny that circumcision is of importance for a 'correct' sexuality, but to suggest, at least in the context of this field work, that the ritual is more than unidimensional, simultaneously incorporating the gestural and the verbal, the physical and the spiritual.

Kalimah

Through ghusl and istibra the body acquires a socially recognized materiality. This recognition is incomplete if the body does not acquire sound. The blowing of the azan in the boy's right ear is succeeded by reciting his formal name. Blowing points in two directions: in the first the boy is initiated into the enunciation of the azan; in the second his name is linked to the verbal intonation present in the official liturgy. In other words, by juxtaposing the name with the liturgy a sympathy is created between the two, at once corporeal and spiritual. Simultaneously blowing gives depth to the body.

It is important to bear in mind that the novice is not the product of the union between a sacred primordial nature and a corporeal nature. If this were the case the boy would be a hypostatized being. The union between the azan and the human form is conceived as one where the latter ingests God. The name makes visible such ingestion. This interpretation is suggested from my conversation with Sadiq Ali on the use of the formal name during the ceremony. For him this use points to a condition in which the novice recognizes the omniscient suzerainty of Allah. The name is crucial in such recognition for Allah has given a name to everything and everybody in this world. I asked him, 'why do you call him Shujat and not, for example, by your name?' 'My name Sadiq points to the way in which this being [pointing to himself] manifests (*zahir*) the lord (*rabb*) in its own peculiar way. It would be sacrilege (haram) for me to call him by my name for otherwise he [meaning his son] will be unable to manifest the lord in his own way'.

The above argument, as I understand it, has two aspects. First, each name manifests the lord to the extent that He has named everything in the universe. Second, each human being, to the extent that he carries a name designated by Allah, is a particularized aspect of that manifestation. The sympathy that occurs between the two aspects is dialogical in that each name is an exemplary indication of the conjunction between the spiritual and the corporeal. What is evident in Sadiq Ali's insight is that the acquisition of sound by the body occurs through the convocation with another voice. In this way, the body acquiring sound learns to be obedient: to listen to the voice is to obey. Henceforth, it is considered natural for the boy to obey his father.

The problem, however, remains: why is the formal name uttered during the ritual? The gestural and verbal inscription on the body are two series regulated by the body. Through the first series of events the body is marked so that it enters into the productive and reproductive life of the domestic group. The evidence is the wound as an eternal truth. The second series reverses the sequential temporal ordering of the body so that through enunciation it enters into the community of Islam. The evidence is the azan written on the body. The azan facilitates the transition from one series to another since the communication between the verbal and the gestural is possible only after the azan is uttered. The name in this scheme guarantees the conjunction, albeit a particularized one, between the corporeal and the spiritual.

The body of the novice thus described is understood through two sets of polarities: the gestural and the graphic and the corporeal and the spiritual. The former shows how certain events are marked on the boy's body, while the second set shows how the body is related to its internal and external environment. Further the first set refers to a series of events and the latter to a series of attributes. The two divide the body into a surface and a depth and in the process show how the body as an object is linked to the regime of signs impressed on it.

The Surface and the Depth

The novice, in terms of a geography of his body, is situated relative to the three realms of a depth, a celestial height and the surface of the body. The depth in the sense of emissions that are polluting and dangerous must forever be controlled. In the heights he finds the word of God writ large. He must always ascend or descend to the surface and in this way claim the new status thrust on him. The body cannot be located in the celestial domain because then it would either lose its corporeality and ability to procreate, or, in the words of Miriam, have the characteristics of an ungrounded alif.

What, then, is the surface of the body? As I understand it, the surface is a kind of frontier available in a series of signs laying down an acceptable and accepted mode of behaviour. These signs, embossed on the body, both through word and gesture, enter into a surface organization which assures the resonance of the series of events and of the attributes. The surface of signs does not, however, imply as yet a unity of direction or community of organs. In terms of a series of events it is primarily the sexuality of the male that is constituted at the surface of the body. The barber's comment that the boy is the penis and without it he is nothing is an apt illustration. But more important, the penis must be made visible and in this way forced into a hygienic sexuality. The ritual distinguishes between the depths of the body, always corrupting and therefore to be guarded against, and the zones of the surface, erogenous but always to be legitimated. All the events of the ritual are, in this sense, coordinated in the genital zone. Here, the phallus does not so much play the role of an organ as of an image which shows the healthy male, thereby pointing to its synechdocal character.

Yet it is recognized that a body will emit both fluid elements (urine, faeces, phlegm) and hard substances (nails, teeth). The elements and substances either emanate from the depth or detach from the surface. Sounds, smells, tastes and temperatures refer to emissions from the depth whereas visual determinations refer to the surface. The relation between the depth and the surface is one where emissions arising from the depth pass through the surface, and as they detach from the body are replaced by a formerly concealed strata. What is important is that such emissions are understood at a distance as being located on and in the physical object of the body in that they are recognized and controlled by another.

If these signs emerge from the physical object of the body and are simultaneously enacted on its surface, it is also evident that they are marks of the presence of the other. The other is neither an object of gaze nor a subject. It is a kind of a priori structure of the possible genealogical positions, conjugal relationships and so on that are potentially available to the boy undergoing circumcision. In this sense the other is a distillation of time by which the rhythms of the body are broken up into units.

What is the relationship between the corporeality of the body and the other? In the first instance, the other in imprisoning the elements and emissions of the body within the limits of verbal representation fabricates bodies out of these elements. In other words, the other in organizing and pacifying the depths of the body, functions as a legitimation of the circumcised body.

There is a second way in which the other is conceived. Through the recitation of the azan and the whispering of the name a theophanic other is sought to be created. This creation arises not so much from the depth of bodies as reside there. The source of this residence is an external environment in that it emerges from a celestial height. Creation itself has two sides. First, its seat of residence (the depth of the body) manifests divinity by which the body becomes transparent. Second, the recitation of the azan is embedded in a corporeality and because of this linked to a particularized apprehension of divinity. Each recitation of prayer in the ritual, in the sense noted above becomes a recurrence of creation. The breath of prayer in the right ear of the circumcised introduces the idea of the guide who stands before the faithful. The other in this sense is the aid. The second part of the recitation — the whispering of the name in the right ear — is based on the attitude of the body prescribed in the course of ritual prayer: erect stance (*qiyam*), inclination (*ruku*) and prostration (*sujud*). The name of man and the prayer of God are thus copresent. Mohammad Umar, a distinguished weaver and the husband of Miriam, mentions that each living being manifests one or all of the three postures given in prayer. The upright stance of the faithful corresponding to Miriam's alif or the celestial height; the lateral movement of animals corresponding to the surface of bodies; the descending movement of plants corresponding to the depth of the body. Khatna, he says, recognizes the three dimensions of prayer in that it has the three postures of the body built into it: erect stance of the mother and child as they are draped over by the green cloth; the descending movement of the child's blood after his prepuce is removed; the burial of the prepuce under the nuptial bed. Each of these three postures is informed by the azan. Circumcision refers to the body in these three dimensions.

What is the body composed of after the ritual has been enacted? In terms of its depth it is constitutive of emissions that are to be purified and controlled. In terms of a celestial height the body is suffused with the word of God. And in terms of a regime of signs playing on its surface the body recognizes the presence of the other. This surface, in the words of the barber, is the communion of the body with its other, evidenced in the first instance through vision (shuhud).

The Everyday Discourse on Circumcision

As we explore the terrain of the ritual of circumcision we find that the procedures characterizing the operation also organize the construction of the ritual. Meaning is structured around the processes involved in inscribing the ritual wound on the body. In this sense an object marks the discourse on the ritual. Contrarily, in this section an object does not fashion the discourse in everyday life. The discourse on circumcision substitutes signs of the real for the real itself in that the body becomes invisible in everyday conversations. Instead, musalmani shows the simulated generation of differences between the circumcised and the non-circumcised. Such differences cement a sense of community and posit pain as defining one's station in life. The ritual wound indicates forbearance. In other words, the referent in the sense of the ritual wound, becomes an ornamental inscription on the sign. This section is concerned with the way the body is absented in everyday discourse and the substitution of khatna by the signs of musalmani.

This is not to suggest an irreconcilable difference between the two terms, khatna and musalmani. An authority links the two terms together and allows for an interchange between them. This authority, drawn from individual and collective memory, makes possible a reversal and a transition into a community. This community is delineated by recalling the ritual as it has been effected on someone else. Furthermore, this recall is founded not so much on an orthodoxy (of texts, ritual practices and formal exegesis) as on the capacity of musalmani to enter into a duplicative relationship with other terms, such as *iman* (belief), azan and so on. In showing the connection between khatna and musalmani I do not postulate two opposed terms whose antinomies are transcended by a third. I assume circumcision is subject to multiple meanings, that plurality in this context is originary.

Everyday Speech and the Disappearance of the Body

As we chart the verbal representations dealing with musalmani we find a fragmented discourse articulated on the heterogeneous practices of the community. These practices are located within the domain of the everyday. They range from the act of weaving to casual social intercourse around a tea stall. The verbal representations of women are absent from the discourse. The body is absented in two ways. The ritual wound is imbued with an incorporeal value while the body is often seen as an appendage of the community.

Musalmani, when it refers to the genital zone, situates the latter within a speech domain where the emphasis is not so much on the physical condition of the body as on an incorporeal valuation of it. The incorporeality of the body emerges from the following conversation immediately after the patwari, mentioned in the beginning of the chapter, had made his comments and left us. The patwari asked his questions in a gathering among which I knew four others: Azeer, Kalimullah, Itrat and Rafiq. The first three, roughly of the same age, were friends of mine. Azeer was my companion on my initial reconnaissance field trips, and in 1985, the only unmarried one among the three. Rafiq is older. He was an outsider in this conversation. After the patwari left us I asked Azeer the meaning of the term musalmani.

A: With musalmani we become Muslim (musalmani *se hum musalman bante hain*).
D.M.: But what is the meaning of musalmani? (*par musalmani ka matlab kya hai?*)

To the amusement of the others Kalim visually mimed the operation of khatna from the perspective of the barber and then asked mockingly,

K: Do you want to become Muslim? (*tumhe musalman banna hai?*).

Ignoring the rhetorical question I asked,

D.M.: What were you before musalmani? (musalmani *ke pehle kya the tum?*)
K: The property of my mother (*Ma ki amanat*).

Itrat and Rafiq nodded assent. Azeer, turning to me, asked,

A: Do you know what masculinity is? (*mardangi malum hai kya hovat?*).

I did not reply. Then Itrat, gently pushing and in an ironic tone opined,

I: What does he know? In the city everyone is adept at masturbation. With musalmani the body acquires strength and we do not have any desire to masturbate. Musalmani and belief are twins. (*Isko kya malum? Shahr men sab apna hi pani nikalat. Musalmani ke sath jism men dum paida hovat aur pani chorne ka koi shauq nahin rahta. Musalmani aur iman jodi hain*).

I asked Itrat,

D.M.: How does the body acquire strength? (*Jism men dum kaise banta hai?*).

I: Strength? I've already told you. We don't masturbate (*Dum? Kahe to diya. Hum pani nahin nikalat*).
D.M.: So? If you don't masturbate you become strong? I'm asking what is the connection between strength and non-masturbation? (*To? Agar ap pani nahin chorin jism men dum ata hai? Men puch raha hun dum aur pani ne chorne men kya taluk?*)
R: The meaning is clear. It is in the nature of things that whenever a Muslim thinks of musalmani his heart overflows with spiritual words. With them the body acquires strength. (*Matlab saf hai. Qudrat ka kamal hai ki jab bhi koi* Musalman musalmani *ke bare men soche to uske dil men ruhani baten ubharne lagte hain. In se jism men dum ata hai*).
K: Go on, you fraud! (*Chal pakhandbaj!*).
R: Are you showing your hand? This hero is going to begin his performance (*Hath dikhariya hai? Yeh lumbardar apni nautanki karega*).

The conversation then examined the difference between the circumcised and the non-circumcised. I will consider this issue later.

An obvious aspect of the conversation is the correspondence established between body and speech. First we find a correspondence between seeing and speaking. Kalim, in miming the operation of khatna, presents his body to the gaze of the other, a gaze where the body making a gesture prompts an understanding contrary to what it indicates. The gesture evokes the sexual organ and ironically reflects on the preceding question: 'but what is the meaning of the term musalmani'. In this sense the gaze consists in dividing the meaning of circumcision. Thus it creates a simulacrum. While an explanation of musalmani is the operation of speech, pantomime is that of the body. This speech, in Rafiq's view, has a spiritual essence in that in his mode of reasoning it is animated with 'ruh'.

However, it is not enough to say that the body is mimicry and speech is spiritual for in the conversation one does not know whether pantomime reasons or reason mimics. There is a complex relation between gaze and speech, for the latter takes on the mode of the former, while the body is effaced under the surface of verbal signs. If sight is ironic so too is speech. Just as Kalim's gestures are interpreted contrary to their indication, so also his question to me ('Do you want to become Muslim?') and Itrat's observation on city folk are ironic reflections.

Such speech reflects on the body in a particular way (Itrat's observations of the body as strength, as belief, as the retention of semen, Azeer's opinion of masculinity and Rafiq's view of the body as spiritual). This speech stands for the body of the circumcised, or more appropriately substitutes the physical sign on the body by verbal signs. This replacement is a relationship of substitution in that the speech of musalmani takes over and selectively arranges those meanings of khatna that evoke the tradition of Islam. In the process this speech suppresses the entire range of meanings available in khatna. This substitution is developed further in the following conversation with Rafiq.[5]

When I met Rafiq after the conversation he felt a need to explain further the meaning of the term musalmani. He began by stating that the patwari was a dangerous man and I needed to be wary of him although what he had said of musalmani was true: you cannot utter the namaz without khatna. More important, khatna does not only make the namaz available for the person in question but also enables this person to carry out his own dua-salam. For Rafiq khatna is the precondition of musalmani. But musalmani and khatna are not the same things. Whereas khatna cleanses the body of impurities, musalmani teaches the person how to approach the Quran and live a pious life. An obvious connection exists between the purification of the body and the conditions necessary for approaching the teachings of the Quran: the latter prescribe an elaborate procedure of purifying the body. However, for Rafiq, khatna is sanctioned only after the Quran has constituted the world in which a Muslim lives. In this sense khatna is chronologically prior to musalmani but, in his view, logically posterior to it. Musalmani designates the way a Muslim lives his life according to Quranic tenets. Khatna is just one, albeit the most important, way of being Muslim.

I asked Rafiq the meaning of 'ruh' in musalmani. In its widest sense musalmani is the recitation of the Quran. Following this it refers to the five pillars. Khatna is linked to the second pillar

[5] Rafiq, one of the members present in the ritual, is Sadiq Ali's father's brother's daughter's husband. Affinally, he is Sadiq Ali's wife's sister's husband's brother. For the purpose of the ceremony he traced his relationship with Sadiq Ali through consanguinity.

(prayer) in that the latter requires ablution. But khatna is more than the purification of the body. It is, Rafiq says, a mark of remembrance that one has heard the Quran and voluntarily recited it (khatna *se* Quran *ka zikr karte hain*). The moment khatna becomes a mark of remembrance in that it recites the Quran it exceeds the body. For Rafiq this excess is the area of musalmani.

In Rafiq's account of khatna and musalmani two moments are discernible. First, khatna is linked to prayer. Second, both khatna and musalmani are subsumed under the power of the recited word. The conjunction of khatna with prayer, while utilizing the liturgy of official prayer, is not merely a public collective act, but also a divine service practised by the fact of being circumcised. The conjunction represents the authentic form of a process of individuation enabling the person to internalize the liturgy.

The inscription of the liturgy on the circumcised keeps the person's body within the limits set by the norms of various hygienic practices, legitimate conjugal relationships and so on. Rafiq's account tells a more fundamental truth. It makes the body spell out the order of musalmani in that khatna is the precondition of musalmani. Khatna in his estimation produces the practitioners of the norm of musalmani. In its most specific occurrence this norm is the inscription of the liturgy on the body and in its most general manifestation it is the recitation of the Quran. To the extent this norm, inscribed on bodies and recounted by them, is repeated in every act of khatna, we find the emergence of a discourse centered around a tradition. From the point of view of musalmani this tradition makes of the body a text which emits signs and functions as a censor which channels and codes bodies. This censoring of the body emerges from the conversation I had with Muhammad Umar after the ceremony of Sadiq Ali's son.

A few days after the ceremony I had shifted to the house of Muhammad Umar who had consented to teach me how to operate the loom. I was a rather poor learner, easily bored by the interminable clapping of shuttles. During one of these sessions Umar remarked that my hand was unsteady because I did not know the discipline of musalmani. First, according to him, the craft of weaving demands the perfectly still body of the weaver. Second, this stillness must be interrupted by regular and abrupt movements of the hands and feet. Such movements are possible only if all the motions of weaving originate from the centre of the body, the loins. For this reason, the loins must be particularly resilient and sturdy in maintaining the still and staccato body. The resilience of the loins is given, in the first instance, through the act of khatna, an act by which one learns the value of pain. The full understanding of pain, Muhammad Umar says, comes from a knowledge of the lives of the *peghambars* of Islam, all of whom endured herculean hardship (the Prophet, Ali, Husain, Ayub Ansari and Sis Ali Salam). To be knowledgeable of their lives is to bear witness. Weavers who are unsteady of hand and slow on their feet are inadequately schooled in the knowledge of pain. Conversely, the strength to bear such pain makes the weaver 'strong of speech' and a 'wise weaver'. Musalmani, he says, is the recognition by the boy of the pain he feels when he is being circumcised. This recognition he then links to weaving in that to be a Muslim is not merely a question of enduring the pain of the circumcision operation, but also one through which the values of resilience, strong speech and wisdom are constituted.

In his statement Muhammad Umar isolates the gesture (the movement of the hands and feet in weaving) to organize his discursive space on musalmani. This gesture maps this space so that its occupants become available for observation and information: whether they are adequate Muslims and weavers. The gesture becomes visible when it shows the inadequacy of the functioning of the working body. A good gesture does not refer to the body but to the positive discourse of musalmani, which talks of pain, wisdom and speech. Thus, a non-discursive gesture is articulated in the language of musalmani. This gesture is a metonymic figure of musalmani, but also a figure where the body of its practitioners are made to speak the truth of musalmani. His statement derives its credibility from what it believes is the indication of pain and its conjunction with weaving.

The Community

In the above conversations we find that musalmani can be expressed and interpreted in various ways. In terms of the informants' accounts it is a statement of belief, the excess of khatna, pain and hardship. Interpretatively, I have suggested that it may be seen as a simulacrum, as substitutive of the ritual, as the truth and censor of the circumcised body. In this sense one may argue that musalmani is productive of multiple meanings. In one respect, however, all the people I talked to were unified in their opinion on the connotation of the term. Each of them maintained that musalmani represented a fundamental difference between Hindus and Muslims.

The dialogue mentioned in the beginning of this section (between Rafiq, Kalim, Itrat, Azeer and me) went on to consider the difference between the circumcised and the non-circumcised. This particular session occurred after Rafiq had left us. Kalim was chided by Itrat for showing disrespect to Rafiq. Kalim launched into a colourful characterization of Rafiq and then commented:

K: What is the connection between musalmani and ruh? Musalmani and pain are twins (Musalmani aur ruh *men kya taluk?* Musalmani *aur dard jodi hain*).

I: 'You speak like a non-believer! There is some depth in Rafiq's statement (*Jahil jaise bolat ho!* Rafiq *ki bat men kuch gherai hai*)'.

K: sarcastically: 'Then make me understand' (*To humko bhi samjha do*).

I: Rafiq said that with musalmani the body learns to recognize pain. The distinctiveness of this recognition is that with it our belief increases (Rafiq *ne kaha ki musalmani ke sath jism dard pehchanne lagat. Is pehchan ki khoobiyat asi hai ki hamara iman badta hai*)'.

K: 'Tell me, do you remember the pain you felt? (*Batao, tumko dard ki yad hai?*)'.

I: 'Not at all. But this is what Rafiq meant. We don't remember our pain because it is part of our belief. That is why musalmani and pain are twins (*Bilkul nahin. Par Rafiq ka yahi matlab tha. Hum dard ka zikr nahin karte kyonke yeh dukh hamara iman hai. Is liye musalmani aur iman jodi hain*)'.

A: 'Through belief we are separate from Hindus (*Iman se hum Hinduon se ilaida hain*)'.

I: 'Yes. But the meaning of the twinning [between musalmani and belief] is also that their [the Hindus'] pain is not productive of spirituality. The Hindu cannot tolerate his own pain. It is true that he can have himself cut in the hospital. But where is the spirit in that operation? (*Ji. Par jodi ka matlab yeh bhi hai ki unke dard se ruh nahin nikalta. Hindu apna dard nahin seh sakat. Yeh such hai ki woh apne ap ko hospital men katwa sakta hai. Par us operation men ruh kahan?*)'.

The conversation ended here. Later Rafiq made the same point. When I asked him whether musalmani distinguished Hindus from Muslims he said:

'It is true that the other word for musalmani is belief which is distinctive to Muslims. This is how we are different from Hindus because for us to be Muslim is to be pure and to have recited the Quran. The Quran is recited with knowledge only after we have removed all that causes impurity in our bodies. The Hindus lack this purity because they are afraid of shedding their own blood'.

Explaining the difference between Hindus and Muslims Umar distinguished between the weavers of the two communities.

'The Hindu weavers here, known as the Kohris, migrated to the city long ago. Sometimes I wish they had stayed because then the nurbaf would have been able to show to them how the weft is made. We are able to weave the weft perfectly straight and without a single knot or break for up to six yards. How is this possible? Musalmani is our secret prayer. Through it we learn not only to educate our hands and feet, but also to maintain the steadfastness of our gaze'. I asked what the gaze was directed at. 'This vision is of the witness who has learnt to recognize pain, to pray with this recognition. Musalmani is the prayer of this pain'.

In each of the conversations musalmani is suffixed by a metonymic and metaphoric progression of meanings. First, musalmani is linked to the recognition of pain, which, in turn, is associated with spirituality and subsequently becomes part of the belief of the group. Finally this series of meanings empowers the body of the Musalman, distinguished from the Hindu in that the pain of the latter is not spiritually elevating. In the second conversation Rafiq connects musalmani to the removal of bodily impurity, links the pure body to the recitation of the Quran and concludes that this progression is the belief of the Musalman. The Hindu, in contrast, lacks bodily purity because he fears shedding his own blood. In the last conversation, Umar considers the Musalman and the nurbaf to be synonyms. The nurbaf is an accomplished weaver because he is witness to pain. Consisting of both gesture and sight, this accomplishment is recognized in prayer. The nurbaf is contrasted to the Hindu Kohri who is a poor weaver because he is inadequately experienced in pain. In each of the conversations the body becomes a metonym of musalmani. However, the full meaning of the term is achieved only when musalmani is linked to general metaphors by which the community is defined.

Common to the three conversations is the ability of each speaker to talk on behalf of the community: the speakers use the plural in talking of musalmani. The community is framed in two ways in the conversations. First, such speech replaces the signs on the body by attaching to musalmani meanings that are external to those found in the ritual. Musalmani is an index of such externality and a way of separation from Hindus.

Second, in the estimation of the speakers the community is represented not only in terms of a separation from Hindus but also in the arrogation of a positive meaning. This is seen in the linking of musalmani to terms such as belief, strength, removal of bodily impurity and so on. Musalmani in this context always exists in the form of a double meaning: the utterance containing or related to musalmani signifies like any other, but also intervenes as an element of metasignification by which the entire utterance acquires a theme. Like a symbol musalmani harbours a double meaning: the obvious meaning both covers and uncovers a figurative one. It is a single signifier which has multiple signifieds. Unlike a symbol, however, musalmani *institutes* a relation between itself and various metaphors. Through this relationship the body is described (and eventually annexed). In other words, musalmani establishes a duplicative relation between itself and other terms such as 'iman', 'ruh', 'dard', and so on. More appropriately, musalmani becomes such in the duplicative relationship it admits with one or other of the terms. This is tantamount to saying that the relationship between musalmani and, for example, iman depends upon the decoder's ability to make the substitutions necessary to pass from one register to another. Yet, the understanding of musalmani is not solipsistic if only because the task of establishing equivalences is already encoded in musalmani. There is something inherent in the term that allows one to read it in a duplicative relation with other terms: musalmani, Muslim, iman, ruh, dard, nurbaf. Whatever the relation between musalmani and other terms, common to the speakers is that musalmani is enclosed by another, the 'Hindu'.

If musalmani enters into a duplicative relationship with terms such as iman, ruh, namaz it is important to bear in mind that the speakers mention pain (dard, dukh) as being the experiential core of the term. All the speakers reflect on this term but in a way that pain is not linked to the physical impairment of the body. The experience of such pain is spatial not because it is restricted to any one body, but because the body of the community, taken as one whole, is a body formed in pain. In the case of Umar the pain borne out of witnessing the hardship of the peghambars of Islam becomes the object of prayer, while for Itrat, in his interpretation of Rafiq's view, pain is belief. Pain, in this sense, is incorporated into the definition of the community, neither disrupting the intentionalities of its members, nor alienating them from the group. For Umar the body of the weaver becomes a presence when it is inadequately experienced in pain.

From the account of the conversations it is possible to make a further inference: the community becomes a presence when the characteristics used to describe it are those by which the body is so delineated. The speakers believe musalmani justifies the pain one feels during the ritual. There is one important way in which this presence is affected. When I asked each speaker whether he remembered his ceremony I was categorically told no one remembers his own ceremony, but that every male at some point in his life is expected to be present at some one else's. Umar maintains it is only by participating in another's ceremony that one understands the significance of khatna for oneself and its linkage with musalmani. Khatna, for him, replicates the physical hardship endured by the icons of Islam while musalmani is the prayer of this pain. Rafiq describes the significance of khatna as one where the person, who is witness to the ceremony, can see how he is part of the community of Islam. In this respect such participation in the ummah arises only after the ritual has been presented to the witness. The significance of this presentation, Rafiq holds, is found in the meaning of musalmani. In its widest sense this meaning is derived from the recitation of the Quran, in its most particular manifestation it refers to the purity of the body signalled by khatna. Itrat holds that the ritual is the first test of hardship for the boy undergoing it. The boy's recognition of this hardship can, however, only be undertaken on his behalf by his male agnates. In turn the boy will understand the meaning of this recognition in his capacity as a witness.

In this respect the speakers base their sense of community on the claim to being members in the community of Islam. In so doing they establish links to that tradition. In effect, their interpretation conceals the work of the ritual that is not connected with pain and belief. An interpretation that reflects on the union of milk and blood and the combination of male and female, on the one side, and the issue of hamdami, on the other, is ignored. It is almost as if that interpretation of the ritual is valorized which, in invoking a tradition constitutes the body as a zone of hygienic practices, but as practices that are commemorative of the community. Both Miriam and Shabnam take recourse to an argument that is legitimated by reference to an inherited tradition and the regeneration of the group. The barber, too, bases her argument on the reproduction of a framed world where the novice eventually recognizes the presence of the other inside his own body. The speakers establish the validity of their case by invoking Islam in one way or another. In this invocation the material properties of the sign, as inscribed on the body of the novice, are effaced. Instead another discourse which talks of belief and pain, replaces the materiality of the embodied sign.

The replacement of the embodied sign by the discourse of musalmani is evident in the constitution of a collective memory. None of the speakers remembers his own ceremony and Rafiq says that it is most significant for the witnesses. In this sense the recall of khatna is based not on its individual inscription on the speaker, but on its distribution among every male member of the community. In describing the common thread that binds one circumcised body to another the speakers attach a retrospective ordering to the ritual. This ordering, evidenced in the discourse of musalmani, transforms individual bodies into a communal body.

Two main operations characterize this memory. The first removes something excessive or adds to the discourse. Second the credibility of this discourse consists in making each body spell out its (the discourse's) code. The act of extracting or adding situates the members of the community within the limits established by the twinning of musalmani with another term. In this respect, the discourse of musalmani is the way by which a social law maintains its hold on bodies and its members, achieved through the disciplinary instrument of the ritual wound. Because this law is incarnated in a physical practice it makes adherents believe that it speaks in the name of the community of Islam.

The credibility of the discourse of musalmani operates through the instrument of the ritual wound in that the latter allows for the linking of musalmani with pain and prayer. Here, the wound is an instrument because it allows living beings to become signs which must recur from one body to another. By situating the wound within its fold the discourse of musalmani incorporates the ritual of khatna, not as a repetition of the ritual in its filigreed detail, but as a community reminding itself of its identity as represented and told by conjoining musalmani with pain and belief. In effect, musalmani commemorates the past as a kind of collective autobiography — a master narrative, more than a story told. It is a cult enacted in its telling.

Conclusion

Bloch (1986) provides a diachronic reading of the circumcision ritual with the intention of showing the principles of change both in the form of the ritual and its constitution as a symbolic system. He says the ritual is historically ordered since it is effected by events external to its framework. However, the '[t]wo images of the world — the ritual and the everyday — cannot in ordinary, non-revolutionary circumstances, compete with each other' (1986: 188) presumably because of the peculiar nature of rituals themselves. The ritual is related to the non-ritual not so much competitively, as in enacting what the latter cannot. As far as the ritual is concerned, the everyday world cannot provide any basis for transcendence. Therefore, the ritual establishes a generalized authority. Allied with this authority is a strong emotional appeal. Together, the ritual's authority and affect explain its individual and collective relevance.

The conclusions of this chapter are almost the opposite of what Bloch shows in his study, especially with regard to the relation between the ritual and the everyday conversations that reflect on it. I do not thereby argue that the ritual and the everyday stand in a relationship of competition, as much as show that the issue is the presence and/or absence of the body of the circumcised. The relation between the ritual and the everyday is one where the latter selectively incorporates those elements that privilege a sense of the community of Islam. Though this community is not identical with that posited in the ritual — the conjunction of the name with the azan — it establishes a continuity with the ritual to the extent that the wound becomes, in everyday conversations, an index of pain and belief. In so doing, such conversations attach a significance to the wound in a way that it is external to the frame of the ritual. This significance marks a separation from the term 'Hindu'. In the ritual this term conveys a state of undifferentiatedness which the ritual then disaggregates. The conversations, on the other hand, value the term so that it offers a counterpoint to what the Muslim is not.

The appropriation of the ritual wound by the conversations raises a second point. It is possible to argue that the ritual, more than the everyday, has a plasticity built into it and is subject to plural interpretations — the accounts of the mother and the barber as well as the valorizing of the body. This is not to deny that the ritual has a transcendental authority, both collective and individual, built into it — the conjunction of the azan with the name which is then mapped on to the body of the novice. But it is not as if such transcendence is lacking in everyday conversations. The sense of community posited by the speakers is based on an apprehension of the ummah, but a community that is established eventually through an act of memory. In contrast, the ritual inscribes the community on the body of the novice.

In the case of this study, then, the basis of separating the ritual from the everyday cannot be the absence or presence of transcendental authority, but the presensing or absenting of the body. In the ritual the body is the object that is worked upon, while in the conversations it is pushed into the background. The ritual, we have seen, not only valorizes the body as the site on which the social structure and the community of Islam are enacted but also brings the body to exist in the elements of language. It is in this sense that the circumcision of every male child is the reproduction of the community.

This reproduction, it must be acknowledged, is initiated through the agency of the ritual wound. With the ritual the body becomes the object of reference and with the everyday conversations the referentiality of the body is substituted by an imagination that constitutes the boundaries of being Muslim. The link between the two — between khatna and musalmani — is provided by the significance attached to the ritual wound. Scarry (1985: 161–326) shows how physical pain is insinuated in the 'making' of the world. In her argument both pain and imagination provide a 'framing identity of man-as-creator within which all other intimate psychological, emotional and somatic events occur . . . ' (1985: 169). The closest approximation to the framing events of pain and imagination is work since it both embodies the physical act and simultaneously imagines a not-yet present object. With work and its artifacts pain and imagination become external to the body. Through such movement the privacy of pain and imagination is shared and sentience becomes social. In this scheme the act of wounding is also the act of creating and may be seen as the intentionality of pain and imagining.

In the making of the world wounding establishes an identifiable relation between the human body and an imagined object. The body here is always attributed with a metaphysics. Scarry says that there are three stages by which the body is so delineated. In the first stage the human body and God's voice occur on separate registers, but the events occurring on one register confirm those occurring on the second. Stage two shows how, with the conflation of wounding and creating, the body is engulfed with the voice and vice-versa. In the last stage creation and wounding are, as categories of action, separated but so that the sentient body is substantiated by material and verbal artifacts (1985: 184).

Following from Scarry's argument we might say that the body as a created object occupies two dimensions: the corporeal and the imagined. The body is corporeal to the extent that it ' . . . has all the sturdiness and vibrancy of presence of the natural world . . . ' (1985: 280). The body is also imagination because in the moment of its making (through, for example, the circumcision ritual) it is embossed with a future. Much of this paper has followed Scarry's argument in the second half of her book. However, rather than argue for a relationship of referentiality between pain and making, I have suggested that the link between wounding and pain institutes the imagination of the term musalmani. With the ritual we find an imagination that is projected on the physical surface of the body. Here, the act of wounding appears as willed and is legitimated since it restores the body to the community. In this legitimation the self-referentiality of the physical body (found, for example, in its emissions) is socially censored. Simultaneously, the wound also constitutes a metaphysical body, but here the act of wounding is produced in accordance with an already ordained world (found in the conjunction between the name and the azan).

In incorporating the wound the everyday conversations do not recreate the ritual body. Infact, they deny it its complete referentiality (the union of male and female). In linking musalmani with pain and prayer the speakers constitute the future and past as unlimited in that the community is not invested with a telos. For the speakers the community must exist for all time and every wound must recreate that existence. Each speaker, it is true, bears the wound within his own body, but the power of the wound and its linking with pain is such that the body is invited into it. In this sense the wound exists before the speaker. He is born to embody it. For this reason, both the ritual and the conversations show how the whole body exchanges its organic will for a social and spiritual one.

Chapter Seven

Inner Voice and Outer Speech: The Life History of Sufi Baba

INTRODUCTION

Any attempt at biographical construction must organize discrete elements into a coherent picture and in so doing impose an order that is fictional.[1] The biography of a mystic — the subject of this chapter — is problematic in another sense. In elaborating the relation between the physical and the metaphysical, the spiritual and the corporeal, between his inner voice and that of the living community to which he belongs, the mystic does not operate within a language framed by theological guarantee which only allows a predictive discourse based on convention. In cases where the mystic draws his inspiration not from writing, but from his actions, the problem is of finding an appropriate mode of exegesis. Such exegesis is problematic because it cannot be fixed through the word, and yet such practice must be formulated as if it were a language. This chapter establishes that such practice has two related dimensions: it is always inscribed on the body of the mystic and it is oriented towards an apprehension of the other. That is to say, the body produces meaning. It is the locus of social communication. An attempt to understand

[1] Such fiction then becomes a pathological 'case history'. One mode of explanation concentrates on the capacity for subversion and resistance embedded in mystical language (see Taussig 1987) while a second studies the life of the mystic under the rubric of 'possession' (see Crapanzano, 1980). De Certeau (1986: 101–15) avoids this type of framing by presenting in his study of Surin a 'portrait in pieces'. In this chapter I will attempt not so much to present a collage of disconnected fragments, as to reconcile a contradiction inherent in the attempt to locate the life of one such mystic within a larger matrix. The contradiction addresses the way Sufi Baba attempts to relate the particularity of his mode of worship to the universality it strives for.

this communication is made by interpreting the life of Sufi Baba[2] as available from his accounts and actions.

Over a period of approximately five months I was a frequent visitor to Sufi Baba's dwelling and a willing companion on various journeys to other villages and the festival of Chahullam.[3] In extended conversations with Sufi Baba as well as in observing his mode of expression on public and private occasions it appears that he thinks of himself as feminine in two ways: he identifies with a certain notion of femininity and the community ascribes an ambiguous gender to him. This chapter explores this doubleness by reading the constitution of Sufi Baba's body as a double-voiced discourse. By a double-voiced discourse I mean that he distinguishes between his private relations conducted in everyday life with 'suffering' women, and his public identification with the weaving community during Chahullam. In the first he communicates by establishing a sense of corporeal identity, while in the second he is a recipient of a collective male notion of female sexuality.

The distinction between the private and the public arises from the difference Sufi Baba effects between his inner voice and his outer speech. Inner voice is to be taken as a condition where the person, in this case Sufi Baba, is the author of signs. Outer speech points to a condition where the person is the subject of signs. In other words, the distinction between the private life of Sufi Baba and his more public and extraordinary self corresponds to a distinction between inner voice and outer speech. Secondly these distinctions are eventually resolved when we consider what constitutes worship for him.

As presented here Sufi Baba's biography is necessarily incomplete, not only because he is (as far as I know) still living, but because he relies so much on the spoken word. Secondly, I have chosen to concentrate on his relations with women and not on other aspects of his life — in addition to being a healer of sorts he is a cycle mechanic and owns some agricultural land. Finally,

---

[2] I refer to Sufi Baba in the masculine gender though he does not think of himself as only male.

[3] Sufi Baba is an important personage in the festival. He is identified as belonging to the weaver's community. Chahullam occurs forty days after Muharram in the month of Tera Tezi. Ansaris refer to this month as Chahullam.

this chapter ends by discussing how Sufi Baba's dual body (the private and the public, the active and passive feminine, the inner and outer) as a single unit is implicated in his mode of worship.

Chronology

My meeting with Sufi Baba was fortuitous. While on my way to a nearby town on my motorcycle I was stopped by another motorcyclist parked on the side of the road. He asked me if I could help him start his vehicle. After the fault was corrected — a dirty spark plug — he insisted that I visit him the next day. The motorcyclist had heard of me from some weavers in the village of Wajidpur where I was doing my fieldwork. If I wanted he could, he said, be of some help since he lived in the same village. I did not visit Sufi Baba the next day, but while in Wajidpur I was told I had insulted him by not going when I said I would. I went to visit him immediately but was told by a male member of his household to come back the next day. Early morning the next day I found him waiting for me.

Sufi Baba does not remember how old he is. He reckons he is in his mid-fifties (this was in 1985). He migrated to Barabanki from a village in Tanda approximately twenty years ago. While he owns some land (he refuses to divulge the exact amount) he says the main source of his livelihood is a cycle-repair shop that he owns, situated some distance away on the Lucknow-Faizabad highway. I have often heard male weavers ironically refer to him as 'cyclewa doctor' (a healer of cycles). This comment is directed not so much on his ability to repair cycles, as it is to his claim of curing ailing women of their diseases. The reasons for his migration are not clear. He says he left Tanda after his wife's death during her fifth pregnancy. In Tanda he worked on a powerloom, but after his wife died he found he had no reason to slave at a living. Leaving his four children — three daughters and one son — in his wife's natal home he decided to settle down in Barabanki, next to his elder brother's household. His elder brother died soon after and his family decided to shift back to Tanda. Unable to bear the loneliness and unwilling to go back he succeeded in enticing his son to live with him. He refuses to say more about his worldly genealogy.

Though silent about his worldly kin relationships, Sufi Baba is more forthcoming on tracing a mythical genealogy. He traces his origins, as do most of the weavers, to Ayub Ansari who lived in Medina at the time of the Prophet. Sufi Baba says that Ayub Ansari was the direct descendant of Sis Ali Salam, the patron saint of the weavers of Barabanki. Ayub Ansari married a woman of low birth and this perhaps explains why the Ansaris are known as Julahas. Ayub Ansari spawned forty sons, who in turn produced four hundred and forty-four sons, all weavers. Some of these weavers (Sufi Baba is unsure of the exact number) migrated to foreign lands, and a few settled in Tanda.

Sufi Baba mentions he is the direct descendant of one such male, but that he (Sufi Baba) was a hopeless weaver because at his birth, his mother's milk dried up and he refused to ingest any other food. In desperation his mother took him to the local *maulvi* who was unable to suggest a cure, but was able to diagnose that Sufi Baba was possessed by a djinn intent on forcing him to swallow his tongue. The maulvi advised the mother to take her child to a *dargah* which specialized in treating such disorders. This dargah, Sufi Baba narrates, houses the spirit of a local pir who had effected miraculous cures on women possessed by the devil (shaitan). The problem was that this dargah treated only women. Sufi Baba's mother succeeded in having him interned here for a month. She promised that if her child was cured she would ensure he dedicate his life to heeding the call of women in distress. Under the healing eye of Murtaza Khan,[4] the keeper of the dargah, the djinn relaxed its hold over him. Within a year he was completely cured.

With the passage of time Sufi Baba was apprenticed as a weaver, though he admits he was a slow learner, not very adept with his hands and feet. He did, however, evince a keen interest in learning about the life of the Prophet and he mentions that his mother would often take him to attend the religious discourse of local priests. Recognizing this interest, his father had him enrolled in a nearby seminary (*madarasa*). Sufi Baba became proficient in religious catechism and was seldom chastised by his instructor on

---

[4] Murtaza Khan is the name of the translator of the *Mufidul Mu'minin*. Sufi Baba is deliberately drawing a connection between the sacred book and his own biography.

his recitation. This learning, he believes, has been of invaluable assistance in his later life.

On her death bed his mother reminded him of the all but forgotten promise she had made to the shrine of the pir and urged him to be true to her word. Her death interrupted his studies — he had been studying in the seminary for approximately four years. His father, however, forced him to work on the loom and by his sixteenth year Sufi Baba was married to the local beauty who was 'fair of face and sharp of tongue'. It was only his mother's promise that kept him from divorcing her. His wife bore him four children. Till her death he never looked at another woman.

His wife's death was the turning point in Sufi Baba's life. He felt she had died because he had not executed his mother's wishes. In Tanda he made a few desultory attempts to keep his mother's word. He does not mention their outcome but it seems they were mainly unsuccessful. For people like him, he observes, it is an impediment to be bound by extensive kin ties. He resolved to leave Tanda and with it his profession as weaver. In his words, he was dead to that community. Before that he made one last journey to the dargah which was not entirely unfruitful, for Sufi Baba succeeded in obtaining a couple of texts on diseases and their cures. These are zealously protected by him and he will allow no one, save his son, access to them.

In Barabanki Sufi Baba chanced on the *Mufidul Mu'minin*. He says it provided a path that was true to the vocation of the weaver. He was impelled to articulate its virtues to the Ansaris and remind them of their noble heritage. What is apparent is that in arrogating this role, he manoeuvred himself into the position of a spokesman of the community. By his accounts, he has achieved remarkable success in being considered a member of the Ansari community of Barabanki. According to him he has complete access to the world of both men and women. As far as women are concerned his contact with them is through a sense of touch and sight. He has often told me that all the energy in his body resides in his hands and eyes. This energy has helped him detect the ailing organs of women sufferers. Although he does not know where this power comes from he is certain it has destroyed all carnal passion in him. The absence of carnality has led Ansari men to repose faith in him while talking of sexual matters.

Inner Voice and Outer Speech • 219

As told to me Sufi Baba's story is marked by a number of significant events. His birth, he takes pains to emphasize, was not an ordinary occurrence. His mother could not produce milk and he was possessed. This possession was cured following the promise of his mother. The second significant event was his interest in religious catechism and his felicity with the word as opposed to the awkwardness of his hands and feet. The third important event in Sufi Baba's life was his apprenticeship as a weaver. The fourth was his marriage and the death of his wife. This was followed by his migration and his 'death' as a weaver. The period between his death and rebirth was mediated by the word. Rebirth here entailed the acquisition of a new body. In what follows, I will discuss this body in detail by looking at his relations with women and his participation in Chahullam.

THE HEALER'S PRACTICE

The Intransitive Body

Seen from within the context of his community Sufi Baba is not, at least initially, striking to look at. He is of medium height (a little taller than sixty-five inches). He has a typical beard, worn by priests and orthodox Muslims of the area, which is streaked with gray and orange (regular use of henna makes it orange in colour). Orthodox Muslims of the area will never colour their beard. He has a thin frame and gaunt face with *kohl* in his sunken piercing eyes. His thick lips are perpetually red with betel leaf and in his mouth one can see only two molars stained with betel leaf juice. His hair is the most significant part of his body. It is long and black (he probably dyes it), reaching up to his shoulder and when he visits women he leaves it untied. On such occasions he wears a little perfume behind his ears and a long flowing robe. His hands are long and slender, perhaps made longer as he does not pare his nails often.

When Sufi Baba visits ailing women he takes special pains to wash. He says this is his namaz (he does not attend the Friday congregation). There is a deliberate attempt at marking his body: black robe, untied hair, perfume. In addition he is, he says, infused with a certain mood which I will discuss later. For this visit he has to be called, he never goes. By this is meant that not only is he formally invited to look at a particular women, but he must feel an inner urge to visit. He calls himself a healer of 'pain'.

On two occasions I was with him when he was summoned by a male to visit women in distress. In the first instance, this invitation was delivered to him not directly by the inviter, but by his own son. On receiving the invitation in the inner recesses of his dwelling Sufi Baba listened keenly to the description of pain which the inviter had provided to his son. The only questions he asked were whether the woman was pregnant and her complexion. She was pregnant and dark complexioned. He insisted that since the invitation was received during the day he could not visit the ailing women immediately, but that he would be by her side after sunset. In the meanwhile she was asked to exercise patience (sabr) and if the pain grew too intense she was to take a required number of dark brown pills.[5] The second invitation was also received during the day. After hearing the cause of complaint — the woman suffered from severe abdominal pains, she was old and not pregnant — Sufi Baba professed his inability to cure her. He was certain she was soon going to die.[6]

Sufi Baba says that during these visits a special juice flows in him — it comes from the heart (*qalb* or *kaleja*) and just as in every emotion, distinguished from sensation, this juice has an inner temperature. The source of this inner temperature is the heart manifested as spirit (*ruh*). The heart also leads to sensory emotion, which, in turn, leads to carnality (*nafs*). The appetite of nafs is lustful but this does not move him when he visits women. The heart is the centre of the body (*jism* or *badan*). To explain the relation between qalb, ruh, nafs and jism, Sufi Baba says the outer body (jism) is a *ghilaf* (a sheath or glove) which successively covers, as it proceeds to the centre of the body carnality (nafs), spirit (ruh) and the heart (qalb).

While visiting women who are in pain he does not feel a lustful emotion because this feeling does not arise from his senses. The inner temperature that moves him does not need an external object in relation to which this feeling is oriented. In other words,

---

[5] Incidentally, he had given me pills similar in size and colour for what he insisted was chronic constipation. He claimed he could determine my ailment from my appearance.

[6] She died a month later of a stomach disorder.

there is a kind of intransitive mental feeling of being in love, prior to this love being pegged on to an object. This object is not given narcissistically in advance, i.e., given in Sufi Baba's own body. He says that the heart is an immaterial covering which, when stripped away leads to direct mystical union('urs) with God. This stripping away is achieved through a process of contemplation which is a precondition for the immortality of the body. This contemplation constitutes worship for him.

When asked why he feels this way only in relation to women he says it is important to contemplate on an object, for without the latter there is the danger of being corrupted by carnality. The secret object of contemplation given to every creature in the presence of the divine is the untouched virgin who cannot be violated. She cannot be violated because for Sufi Baba while contemplating the divine one must be completely pure in one's heart. The virgin comes closest to this notion of purity. He prescribes three steps by which this object can be apprehended: reason (aql), memory (yad or zikr) and knowledge of sacred texts ('ilm). Of these the most important is memory for it shows the relationship between tongue (zaban) and word (kalam). In addition to feeling and contemplation, Sufi Baba explicitly constitutes his body through movement. This movement, as far as his relationship with suffering women is concerned, is visual and tactile. He mentions that the power of his gaze and the movement of his hands over the affected organs helps him locate the source of pain.

In his relationship with women it is apparent that Sufi Baba constitutes his body as feeling and contemplation. While feeling arises from the mysteries of the working of the heart, contemplation leads to the attempt at unveiling these mysteries and is eventually anchored to an object. This object is internal to the body but not of the body. By this is meant that while the inviolable virgin is embedded within every living creature she exists only so that everybody can achieve nuptial union with God. In this sense everybody is female while God is male.

To the extent that Sufi Baba's field of vision has the final effect of emphasizing the body — his own and that of the patient — it may be argued that through contemplation he consecrates the body. It is important to point out that he does not advocate a subjectivity imprisoned within its own thrall precisely because he is able to step out of himself through his mode of contemplation, and by touching the affected organs he is able to internalise this stepping out. That is to say, his sense of touch and sight extends to tactile perceptions. Through sight he is able to recognize the autonomy of his patients, while through touch he becomes an other to himself.

Therapeutics

There is an elaborate procedure before Sufi Baba begins his cure. This procedure can be discussed only on the basis of Sufi Baba's description as weavers are reticent in talking about him. As mentioned earlier, he visits women after sunset. After he has readied himself he enters the house of his patient through a door directly leading into the zanana. If he must enter through the mardana he insists that male members be absent. Once in the zanana, unless a helper is required, Sufi Baba tries to ensure no one is present while he is seeing his patient. The conversation between patient and healer is usually one-sided: the patient does most of the talking. Sufi Baba holds that verbal intonation reveals the extent and depth of pain.

Pain or illness/disease is divided into two broad categories: the first raises the temperature of the body while the second lowers it. A proper balance must be struck between these two states. This balance deals with the relation between nafs and ruh, for in Sufi Baba's estimation, pain arising from a heating of the body results from an excess of sensuality, while that which cools it is a consequence of excessive spirituality. Accordingly the body is divided into a spiritual and sensual zone and the cure for pain is found depending on the zone that is affected. There are four such zones: (1) the contemplative zone i.e., the heart, (2) all orificial areas of the body — nose, mouth, eyes, ears, anus and vagina, (3) the doing domain, the hands and the feet and (4) the reproductive field, i.e., the upper (nurturance) and the lower (procreative). Each of these zones is divided between sensuality and spirituality. For zones two, three and four the division between spirituality and sensuality corresponds to the division between the upper and the lower portion of the body. As far as the heart is concerned, it is divided equally between ruh and nafs.

In turn each of these zones is identified with the four basic elements — earth, water, fire and air. The lower portion of the body is composed of earth and water, while the upper comprises of fire and air. Thus while the nose, mouth, eyes and ears have a preponderance of fire and air, the anus and the vagina are constituted mainly of earth and water. Similarly, the hands are concerned with fire and air, and the feet with tasks dealing with earth and water. As far as the heart is concerned, because it is constitutive of spirituality and sensuality, it is divided between earth and water on the one hand, and fire and air on the other.

The cure for the suffering body depends upon a balance being struck between spirituality and sensuality and thus restoring to the body the proper distribution of the elements. The cure is based on treating heat with cooling elements and vice-versa — fever is to be treated with elements composed of fire and air. Sufi Baba mentions that when a woman is suffering from high fever — gauged by the feel of the pulse and the colour of the skin — he first gently blows his breath on her and then gives her pills that have been roasted over a slow fire. This fire exorcizes the medicine of whatever heat it has. The pills are then allowed to cool at room temperature. The idea is that air will reinvigorate the pills with spirituality.[7] Sufi Baba says that for illness arising from excessive spirituality he recommends oil massages. He makes the oil at home. The mixture by which this oil is made is a closely guarded secret. The oil massage helps in reinvigorating the body and returning it to its former balanced stage.

In addition to cures dealing with the impairment of the outer body, Sufi Baba says he is proficient in confronting the 'pain of the heart' (*dil ka dard*). This pain arises from a confusion between spirituality and sensuality and leads to the impairment of the contemplative faculty of the sufferer. These are the most difficult cures to affect for, as he puts it, 'we have to become partners in pain' (*dard men sharik*). In taking on the pain of the sufferer Sufi Baba provides a cure. In taking on the impairment of his patient, Sufi Baba admits there is the danger of him becoming impaired.

[7] Once when I was suffering from high fever he blew on me and gave me the ubiquitous brown pill. When the fever showed no signs of abating, he withdrew his treatment not because I was male, but because I had in his estimation been corrupted by the big city. On another occasion after I had hurt my rib he applied a poultice to the affected area. It provided almost instant relief.

His asset is that he can retain the power of his contemplative faculty precisely because he can see more. In taking on the patient's pain he says he wills a sympathetic pain in his own body. This allows him to respond to his patient by complementing her experience with his. He often says he provides sustenance to the woman by teaching her the value of patience.

This impairment which distorts the contemplative faculty of the patient, is, in Sufi Baba's opinion, usually the result of possession by a malevolent djinn. In cases of extreme possession, he admits he is unable to find a cure because he has not been formally tutored. On the other hand, his personal experience gives to him an immediate empathy with the woman and because of this reason, he can see deeper into her body. While unable to prescribe a cure under cases of extreme possession, primarily due to inadequate knowledge, he can diagnose the nature of possession.

What is important in the prescription of cure is the way the body is constituted. In his relationship with women Sufi Baba relies on touch and sight to diagnose a cure. The cure rests on visual and tactile sensations: the apprehension of impairment becomes possible only through the movement of his gaze and hand. This power of sensing and acting is derived from his own experiences — he is able to share the pain of his patient on the basis of the reaction of his own body. In this sense, memory is inherent in the power of sensing and acting. This memory is a knowledge of the movement of the gaze and the hand, and second, a recognition of what this movement stands for. While the first is a habit of experiences of similar debilities, and thus necessarily located in the past, the second is imbued with new intentionality: only on the basis of such movement is a cure administered.

This intentionality is an active interest that permeates his feelings and touches him the most closely. Because this feeling is intentional it refers to the mode of the future, to the not yet. In this intentionality though each of his clients is a distinct individual and the cure varies accordingly, he claims it is his active intervention in the pain of the patient that allows for the diagnosis. This active intervention is 'read' in terms of his sensory perception of impairment, but the urge to cure is prior to feeling. Though in a formal sense this reading is based on a classification of the body, it is left to him to interpret signs of impairment.

This interpretation, he says, is eventually a result of his contemplation of the divine.

His intentionality must result in a sphere of certitudes for otherwise a cure could not be prescribed. In turn, these certitudes stem from the structure of the experiences of previous cures as they have been embodied by him. While Sufi Baba's experiences of curing women sufferers is based on his knowledge and recognition of bodily impairment, it must not be forgotten that this knowledge and recognition, seen dramatically in the case of possession, is always embodied by him. Thus, while the prescription of cures can be transferred from one patient to another, the experience of impairment which he says is embedded in him cannot. Crucially, the history of cure, as conceived by him, is marked on his body as much as it is available in the accounts of his patients. The inscription on the body of a therapeutics of pain points to a conception in which his 'body is memory', in the sense of de Certeau (1986: 227). In this way he is the author of signs and because no other body is privy to his experiences, he constitutes his body intransitively.

It should be added that Sufi Baba's therapeutics, whether it diagnoses contemplative pain or physical impairment, has the final effect of emphasizing his own body. Meaning is produced by how his own body reacts to the suffering of the other. To the extent that we can trace his method of curing to his object of contemplation (the virgin), it can be argued that the other is not a particular but a signifier that produces infinite meaning. In other words, the impossibility of a limited meaning valorises a condition where he is altered by the other. He is certified by his alienation.

Chahullam

A second way by which Sufi Baba constitutes his body is seen in his relationship with Ansari men during the festival of Chahullam. Here, Sufi Baba does not author signs. Instead he is a member of a community that privileges him with playing the role of a mediator between males of this community and other inhabitants of the area and between Ansari men and prostitutes from neighbouring towns who sell their services during the festival. Because Sufi Baba is formulated within a determinate role given to him by the weavers' community, he is the subject of signs. In his mediation between weavers and other inhabitants these signs are both embodied by him and stand for the community. In this sense his body can be seen as transitive. To the extent that he mediates between weavers and prostitutes he encapsulates the idea of an androgynous body. I will discuss first the festival.

During the month of *Tera Tezi* or Chahullam, taziyas are ritually interred on the banks of a river which runs just beyond the southern section of the village where this festival occurs. Among Muslims of the area Tera Tezi refers to the thirteenth day of the month, a day before the appearance of the full moon. For weavers Chahullam is perhaps the busiest month of the year. It is not only that they make elaborate taziyas, but virtually every Ansari hearth of the village has a few guests staying in it who have come from both within and outside the district of Barabanki. Among all Muslims identified with particular occupations, this month signifies a lack of restraint, evident in the behavior of young Muslim women in public places and occasions. The veil is disposed off, bright clothes are worn and verbal interaction with Muslim males of approximately the same age group frequently borders on the vulgar.

Chahullam may be divided into two broad types of activity. The first, stretching from sunrise to sunset, is characterized by the procession of taziyas, the beat of drums, songs and dances and the eventual ritual burial of taziyas. The second type of activity begins after the burial when the sun has set and continues well into the morning of the next day. It must end before the sun rises. The second part of Chahullam is a time when the order of the day is subverted. This subversion is expressed both by privileging the sensual and mocking the spiritual zones of the body. The image is of the body that fecundates and is fecundated. The difference between the two activities is that during the time of day the body is not mocked. For both day and night the image of the body is never individual, but may be seen as the inexhaustible vessel of death, conception and reproduction. This is the time of the *mela* where a number of things happen simultaneously — nautankis, wrestling duels, poetry sessions, hawkers selling *biryani*, kababs and various sweets. The peripheries of the mela are centres of furious activity, for this is where the *ganja addas* (marijuana dens) are located, and beyond them the tents of prostitutes. In the year this fieldwork was conducted, the local politician, who was also a member of the Legislative Assembly and a high caste Muslim had banned prostitutes and marijuana from the mela. He was not entirely successful. In each of these activities Sufi Baba occupies a central role. In the first, he heads the taziya of the biraderi, while in the second, he extols the virtues of individual prostitutes and customers alike.

From the third day of the month of Chahullam preparation for making taziyas among the Ansaris begins in earnest. Each Ansari group makes its taziyas and keeps it hidden until the thirteenth day in the house of an elder member of the group. The community as a whole also makes a taziya which is the most spectacular of them all. Till the period of their storage taziyas are covered by a diaphanous white veil. The *mujawar*, who maintains the taziya and will lead the procession when it is to be ritually buried, ensures the secrecy of the taziya before its public unveiling.

Taziyas, made of thin wooden strips and paper, are a riot of colour with pink and gold predominating. Standing upright they measure from between six to twelve feet. The main taziya has four sides to it representing the local mosque, the Taj Mahal, the Delhi Jama Masjid and the Kaaba in Mecca. Craftsmanship is intricate and it is not unusual to find expatriates from West Asia constructing taziyas.

The Transitive Body

The construction of taziyas corresponds with heightened public activity in the night. Soon after work has begun on the fourth day, the Ansaris take out their big drums (*nakkaras*) after the evening prayer and begin to dance to their playing. Several groups of men first dance around the Ansari quarter of the village and then the local mosque. Each group has males dressed as females who will be the inevitable centre of attraction. Most participating women are old. Young women congregate around the local mosque with trays of sweets. Men say that in this way their women, particularly barren wives, are preparing for asking for boons (*mannat*) during the thirteenth day, when taziyas are put on display.

In these festivities Sufi Baba has a prominent role to play and is usually the centre of attraction when he is performing. Here, he dresses as a kind of androgynous clown. He wears clothes that are half male and female — a lehnga and a kurta. Simultaneously mocked and mocking, he is seen as someone who, because of his natural affinity, is able to dissolve sexual categories and boundaries. This is quite evident not only in his visual appearance but also in the conversation occurring between him and other members of the dance group. A kind of informal scatological question-answer session emphasises the visceral areas of the body.

Q: Tell us Sufi Baba about the marks your husband left upon your breasts?
A: They were made by the female barber when she was removing my hair on our nuptial night.[8]
Q: I hear you can accommodate ten adult men all at once.
A: Open my door and see for yourself. In any case the ten men were tied too tight by the thread.[9]
Q: I don't see your breasts Sufi. Last year they were over-ripe water melons.
A: After the vegetable vendor[10] was through with them all that remains are cow's teats. This year I will offer you my shit.

The question-answer session is not a simple form of play. It has a dark side. Sufi Baba has to maintain the masquerade and struggle for self-control in circumstances that are often humiliating. Inevitably, he is questioned about the true purpose of his relationship with women and he has to make an effort in such instances not to insult and yet appear witty. It is almost as if the ritual context of this drama requires Sufi Baba to be in a state of intense self alienation from his avowed vocation as healer of women and contemplator of the divine. He does not, he says, feel insulted because the questions are often lewd, but because there is no relationship between tongue (zaban) and word (kalam). The point is that he seeks to establish an identity with women on the basis of suffering.[11]

[8] The reference is to an upper-caste Muslim woman who, two weeks before her marriage has, according to a belief of the Ansaris, her hair assiduously removed from her body by a female barber.

[9] The reference is to the Hindu *pundit* who is thought to be extremely timid in sexual matters because he is always constrained by his sacred thread.

[10] The *Kabariya* or vegetable vendor is acknowledged for his stinginess and his legendary capacity of extracting the smallest profit whenever possible.

[11] In his study of Al Hallaj, Massignon emphasizes the solidarity between social suffering and the healing pain of salvation (see Massignon 1982, vol 3: 111–21).

More important, the question-answer session accentuates a relationship of antinomy between him and the community. It almost appears as if his performance has perhaps only subjective coherence. However, to the extent the community of questioners is itself the other in relation to his delirium, his performance is not something that makes sense only to him. On the one hand, Sufi Baba's retorts are authorized by the audience, on the other, the audience tends to control his answers by allowing only interpreted or corrupt versions to circulate. In other words, the community assigns a place to him, but a circumscribed one. By situating his discourse within itself, the community takes charge of it, limiting his discourse to a truth known on the inside, while allowing another on the outside, one that is noble and formal and found in the *Mufidul Mu'minin*. On his part Sufi Baba is not a passive recipient of this attempt at authorization. He, too, is engaged in a politics of utterance, not because he is engaged in the will to persuade, but because he has a need to be heard. In this sense he attempts to create a public place for communication where the erogenous areas of the feminine body are recognized, albeit through mockery.

The destruction of Sufi Baba's dignity is accompanied by the theatricalization of his body. This 'stigmatised self' is the locus of decomposition and breakdown where 'faith' arises. De Certeau (1986: 43–44) gives the example of St John on the Cross where the mystic voyage, from start to finish, is characterized by a series of negations in an attempt to reach 'that which exceeds'. That is to say, each stage arises out of the non-identity of the subject to the state which he is in. Perception, vision, ecstasy and putrescence are cut off so that the discourse of St John on the Cross is an indefinite series of not it, not it, not it. In the case of Sufi Baba this negation, understood in the first instance through the inscription of the other and an authorial voice, is also marked by the discourse of his community.

Together with his appearance and ability at repartee Sufi Baba does a curious dance. He swings his hips and pouts, trying to simulate, it appears to me, a cabaret dancer in a Hindi film. Despite the delight of spectators in this masquerade, his performance is not all joyous. He tells me his comedies are played out against a background of exile, personal loss and continuous personal trial. This exile is expressed not merely in terms of a loss of his kinship community, but also because he does not belong to any Ansari group within Barabanki. Eventually, because he does not offer prayers in the way of other Muslims, his exile is the loss of an external prophecy. His personal loss, similarly, is not merely the death of his wife, but also the loss of an explicit gendered identity. Finally, he feels he is on continuous personal trial, not only because his cures must be effective and must be seen to be so, but because in his contemplation of the divine he has to overcome numerous obstacles which arise as a result of his senses.

The Spokesman

Sufi Baba's participation in the festival begins in earnest from the eighth night of Chahullam. The centre of congregation is the local mosque. The nakkara procession moves out of the Ansari hamlet and stops for a while before select individual hearths of both Hindus and Muslims. With each succeeding day until the thirteenth, the pace of playing the drum and of dancing and talking into the early hours of the morning is maintained, eventually reaching its peak on the twelfth day, when various Ansari processions remain active through the night. Meanwhile the community taziya is given its final shape and form.

After the first call of the thirteenth morning the community taziya is unveiled near the local mosque. The procession of taziyas begins with drum beating. Sufi Baba, dressed as a male, edges to the front of the procession. His place there is not without tension. Some members urge him to stay at the rear of the procession while others wish him to remain where he is because of the felicity of his tongue. In its movement the community taziya circles the Ansari quarter and stops before the dwelling of select members of the community. Usually women of these dwellings emerge from their houses to ask the mujawar questions about the future. Often Sufi Baba interjects and supplies answers. He says the answers given must not be immediately transparent but should provide the questioner food for thought.

Q: My daughter needs a husband. When will she get married?
A: Muhammad Ayub had forty sons and four hundred and forty four grandsons. The name of your daughter's husband

is written in the progeny of one of these elevated souls. He will come to her not as a stranger (ajnabi) but with the fruit of his ancestors.[12]

Q: Our community has a hallowed name. When will the weaver gain recognition for the nobility of his craft?
A: Follow the kitab and take the name of Sis Ali Salam in your daily actions. God willing the time is not far when others will look at us with envy.
Q: Why is it that the biraderi does no help its members who are in need? [This question is invariably asked by women].
A: Those who turn a blind eye are not the sons of Adam.

In sharp contrast to the laughter of the carnival this sort of conversation seeks to reproduce the self-definition of the community. It signifies in the way a command is issued and is therefore injunctive. In relating what is happening or what has happened this dialogue constitutes the actual by pretending to be the representation of a past happening. That is to say, it assumes authority by appropriating a tradition. For this dialogue to establish its base of power the procession of taziyas provides a guarantee to spectators that it is a discourse of an actual happening located in the past and in articulating this discourse the Ansaris are true Muslims. Simultaneously, the answers of both the mujawar and Sufi Baba gloss over fissures within the Ansari community. Representations of the past are characterized by an authority precisely because they obliterate the memory of the conditions militating against them. Further, this dialogue has pragmatic efficacy in that it makes believable what it says. In this sense it is performative.

The Mediator

The second half of the festival is different from the formal ceremony of the first. For Sufi Baba this contrast is not merely between conventional enactments of the day and those of the night, but rather between the limiting events of the day, when the community seeks to define itself, and the playful seriousness of the night, gauged most acutely in the quarter where prostitutes have pitched their tents. The mood here is subjunctive, not

[12] What seems to be said is that this woman's daughter will marry a member of the community. This member is already known to the family.

injunctive. In the first he attempts to formally constitute the self-definition of the community by reiterating the nobility of the weaver's profession as well as his steadfast adherence to Islam. In the second he mediates between customer and prostitute by extolling the accomplishments of one to the other. There is a discrepancy in the meaning of the mediation he arrogates for himself and that granted to him by members of his community. His participation, for him, is linked not merely to the satisfaction of carnality but also to a sense of spiritual love. The customers when they talked to me covered their embarrassment by saying they were seduced by him.[13] I will elaborate this section by discussing the events structured around the prostitutes' quarter.

After the burial of taziyas — characterized by a procession from the centre of the village to its southern end — the focus of attention shifts back to the main square. The nautanki is performed here. The performance and audience are ringed by hawkers selling their wares. Beyond that, as we proceed westward we find the marijuana dens, and still further west, the temporary quarters of prostitutes that occupy a wooded area. Sufi Baba situates himself between the marijuana dens and the prostitutes' tents. He makes a deliberate attempt to distinguish himself from others by dressing differently. He wears the same clothes as he did while visiting women in distress. In a later conversation he told me this makes his intentions clear to prostitutes and enables them to repose trust in him. The dress serves as a marker to signify a particular intentionality. This intentionality is defined by an imaginative identification and is characterized, it seems to me, by an intense emotional state where various contradictory feelings appear to coalesce — of sorrow, joy and pain. The only detachment that comes to him is one which accompanies his comic aspect, but even this act is shadowed by anxiety in that he is driven towards renouncing his sexual desire.

Sufi Baba says that well before Chahullam some members of his community ask him to reserve women for them on the thirteenth night. Some of them find it difficult to solicit the

[13] My account is gathered primarily from Sufi Baba's description. I could not talk to any of the prostitutes despite Sufi Baba's unilateral attempts to find an appropriate woman for me. My conversation with males occurred well after the festival. I was present on two occasions when Sufi Baba tried to solicit customers.

services of prostitutes, while others express a liking for idiosyncratic sexual postures and find it embarrassing to convince women to participate in their desires. Sufi Baba in such cases is obliged to find the most appropriate combination. A third type of male needs no mediation and is one who often makes the mediator the object of his ridicule. On one occasion when I was present Sufi Baba made his pitch with the following prefatory remarks:

I have this necklace to sell
Hema Malini took it off and became fat and ugly.
Mumtaz wore it and sent God to hell.
Make her wear it, she will become a houri.

On getting no response: 'Come, come, the Mumtaz of Mughal-e-Azam anxiously awaits the Qutb Minar'. A further lack of response: 'Go away you stinking goat, you have left it implanted in that she-goat wife of yours'. Saying this, Sufi Baba laughed, shook his hips and signaled the man to come stand by him. In a high falsetto voice he told the man that the latter had been enslaved by his wife and if he was not careful he would become impotent. In a later conversation he explained that wives are taught to be sexually docile while prostitutes are known to be aggressive. Continued sexual intercourse with one's wife makes the man docile and eventually impotent. It is therefore important to renew sexual ardour by periodic visits to prostitutes.

On a second occasion I was directly involved in the conversation between Sufi Baba and a customer. He had already struck a deal with the customer sometime earlier in the day. On seeing me approaching the two of them the man hesitated, indicating he was there out of curiosity. Sufi Baba assured him I was waiting for a particular prostitute who specialized in a position in which the woman is astride the man (he called it the vagina astride the *minar*). Feeling embarrassed I attempted to distance myself from the conversation. Sufi Baba and the customer insisted I stay. Sufi Baba questioned us on our sexual experiences and then expatiated at length on the various sexual positions one could employ with prostitutes. Often he would demonstrate a position through body gestures, but always from the perspective of the woman. By now a sizable crowd of males had gathered to see the performance. Someone asked him how he knew so much without any experience of sexual intercourse with prostitutes. Resisting personalisation he replied that the prostitute is both a concrete figure and an object of love. He responded to her in terms of the latter. Another voice asked:

Q: If you can't stick it in her how can you love her?
A: If you can't copulate with your wife do you stop loving her?
Q: I'm asking you who do you fornicate with? I see you dressed as a woman, but that thing hanging between your legs is not of a woman.

All the while Sufi Baba continued with demonstrating various postures. On hearing the above question he stopped, asking, 'Can you do what I'm doing? The thing that you point at is passive. I have taught it to be so.' Then he recounted a dream:

Once late at night, unable to sleep, I was thinking of a beautiful woman I knew in my former life. Engrossed in my thoughts I saw a girl approach me who was so enchanting that not for a moment did my gaze wander from her face. The magic of her glance and the eloquence of her words convinced me she was a figure of pure light. She spoke to me of the places she had been to and of the gardens she had visited, and of all the manifestations of God's will on earth, of her house peopled by those in search of divine inspiration. She told me that all signs of love are an indication of His beneficence. I am His prostitute.

Speaking of this incident later he provided variations of his story. The girl of his dreams had also, it seems, told him she had seen people without reproductive organs who had the ability to procreate. The prostitute was one such. When I asked him how this was possible he said the prostitute creates passion in others — she herself is not the subject of desires. For this reason she is the creator. As for his relation with the prostitute he said that in his various manifestations, as healer and mediator, it was to her he referred, and through her to divine inspiration.

Talking to Ansari men about Sufi Baba's role during the night of Chahullam I was given various interpretations. One view held he was a charlatan whose main object was to slake his own sexual desire. At best he dissimulated a sensual love so that he could preserve his reputation for austerity and piety. A second view believed he typified the characteristics of a khidmatgar, engrossed in providing disinterested service to other males. The third, and from my point, most interesting opinion felt he had actually succeeded in renouncing his sexuality and understood women better than others because he was exposed to varied situations. When I confronted him with the three interpretations he distinguished between three types of contemplation: (i) a mode where the attempt is to satisfy one's own desire (the term used was shauq) without any regard for the beloved; (ii) a mode where one's desires (*ruhani* shauq) are satisfied so long as the beloved is satisfied; (iii) a mode where one's desire (*ilahi* shauq) is satisfied through a recognition that the beloved is none other than a manifestation of divine will. This classification contains its own motivation. Love considered in relation to the prostitute is different from love in relation to God who is both the lover and the beloved. The problem is of reconciling spiritual love with physical love. I will return to this point later. Sufi Baba says most people understand love in terms of the first mode, and this perhaps explains their inability to understand him. Elaborating further he says, because the prostitute manifests God's will, he (Sufi Baba) is a prostitute while contemplating God.

There can be various ways of reading the above ethnography. The most obvious is to see it as the product of an ethnographic interaction between the field worker and his informant. Crapanzano (1980: 140) reminds us that the ethnographic 'encounter' never ends, demanding as it does perpetual exegesis. In writing of the encounter time is frozen: the act of writing and the actual interaction are seen as united in their occurrence. Perhaps, the most important character of the encounter is the recognition of the other's subjectivity and the importance of this recognition in one's own self-constitution. In this sense the act of writing is also one of expiation. I, however, have not discussed the life of Sufi Baba as an ethnographic encounter. It is not only that all such encounters are necessarily prefaced by the way the individual, so described, substantiates his picture of the world he inhabits, but also that in reconstructing this world, the anthropologist as fieldworker makes a double incision. In the first sense, the anthropologist enters into the life of his informant during the encounter. In the second, he re-encounters this life in the act of writing. What seems crucial in both these movements is the attempt to constitute the other as 'informant', 'native', etc. This constitution, however, is not the result of a dyadic interaction between the anthropologist and his informant, but involves an interaction within a space constituted through an act of imagination. In what follows I will attempt to mark out the distinctive contours of this otherness through the operation of a certain imagination.

By imagination I refer neither to fantasy, nor a faculty of mind which produces images identified with the unreal, nor also to aesthetic creation. Following Castoriadis (1987) I take imagination to refer to the irreducible capacity of evoking images. While these images cannot be decomposed further, they rest on the symbolic domain to express themselves, as well as attempt to satisfy determinate needs. In effect, the order of the imaginary has a symbolic and functional dimension, but it cannot be reduced to either of them. In the context of Sufi Baba as healer and spokesman there are two ways by which this imagination is constituted: as an active and passive imagination. The healer and the spokesman, then, are linked in the notion of the prostitute, who is both active and passive, reconciling the two aspects of the imagination.

In his relationship with women Sufi Baba constitutes his body intransitively. What is privileged here is a meditative faculty, capable of new intentionality and thus of new meanings. This intentionality is necessary since only after Sufi Baba has intervened in the pain of the woman can a cure be prescribed. This prescription is made possible in two ways. First, pain, perceived through an imaginative identification with the person so impaired and given in the prior classification of the body, is reflected in a sensation of breakdown. Second, pain must be translated into a visionary perception of an ideal body where the four elements of nature are in perfect harmony. Such translation is achieved via a process of contemplation where memory is privileged. The active intentionality of finding a cure takes recourse to a gestural medium in that physical movements are given transiently in relation to an external object (the woman's body). The translation of pain occurs in a graphic medium in that permanent signs, having their origin in these movements are subject to a syntax given independently of any physical interpretation.

Pain itself is divided into two broad categories: physical and mental. Sufi Baba calls the latter 'pain of the heart' and the former, the impairment of the outer body. Physical pain is expressed in phonics, facial expression and the modulation of the voice. In this sense the signals of such pain are communicative and their cure is based on his experience with earlier similar pain. The cure for bodily pain is experiential. Mental pain or pain of the heart impairs the contemplative faculty of the person in pain. Its cure is based on an imaginative identification with the patient: Sufi Baba must become a partner in pain. Unlike bodily pain this pain affects the entire disposition of the person. Sufi Baba is able to prescribe a cure by 'reading' signs of impairment, not on the basis of former experiences, but on how it marks his own disposition according to his classification of a balanced body. In this sense his cure is based on his ability to retain and decipher his contemplative faculty. The cure for such pain, then, is contemplative.

The manifestation of pain is a necessary condition so that a patient's memories might have therapeutic value and so that Sufi Baba's interpretation might have contemplative value. The technique of cure is based on how the patient expresses pain. The cure fails if it cannot accomplish this. Second, Sufi Baba believes that for the cure to be effective the speaking subject's place is decisive. This method elucidates language as an intersubjective practice by allowing a space for the other. I will return to this point.

In his relationship with the community of men in Chahullam Sufi Baba's body is constituted transitively. There is an embodied transference of meaning through parody. Thus, as far as men are concerned, Sufi Baba's body is seen as being marked in a grotesque way with the signs of femininity. This imagination reproduces the woman's body and Sufi Baba is seen as representing the reproduction. It is not only that he dresses as an androgynous clown, but that he is questioned only about the feminine aspect of this androgyny. The parody here extends to his relationship with women where the spectre of dismembered limbs (marks left on the breasts, etc.) impeaches the entire personhood of Sufi Baba. This imagination of mockery is represented in visible form and must be rooted in the immediate, the tangible, the concrete. The flexibility he arrogates for himself is formulated in reference to an exile.

Sufi Baba's performance during Chahullam is informed by a sense of exile in that he is forced to adopt a stance of intense self alienation from his avowed purpose as healer of women in distress and mediator between customer and prostitute. For Sufi Baba the exile is made all the more poignant because in all the questions directed at him the profane takes possession of the word. For this reason the profaning word is incapable of expressing the sacred, and of the norm of the sacred of which it professes to be a judge. For the questioner the ordeal of exile no longer exists. What remains is an imagination whose products are merely passive, in that they are already given. The exile, as he understands it, is an inner exile written in the hearts of those who are exiled from their beloved. For this reason he does not consider his exile so much in terms of a material filiation or an external causality, as much as an exile from his true vocation of realizing his beloved.

In attempting to realize his beloved he employs a personal dialectic given in the two objects of contemplation: the untouched virgin and the prostitute. Contemplation in both unfolds in three stages. For the untouched virgin Sufi Baba employs reason, memory and knowledge of sacred texts. Of these memory is the most privileged since it is the precondition of introducing new meaning. As far as the figure of the prostitute is concerned he distinguishes between three modes of contemplation: the satisfaction of one's own desires with complete disregard for the other; the satisfaction of the desires of the other; a satisfaction of one's own desires through the recognition that the other is a manifestation of God's will. Of these, the last is the most important since it reestablishes communication between heaven and earth. In this sense contemplation is dedicated to a vision of divine things.

Common to both forms of contemplation is the idea of love: spiritual in the case of the virgin and carnal in that of the prostitute. That is to say, love is both spiritual and creatural. As I understand him Sufi Baba attempts to restore a sympathy between the spiritual and physical by answering the question: through whom do we love Him? The virgin and the prostitute are subjects of this love. Such love is physical since it contemplates a concrete image, and spiritual for the attempt is not to possess the virgin or the prostitute, but to embed them within himself. This is what he calls his raz-o-niyaz (the secret prayer of God). In talking of his mode of contemplation Sufi Baba's discourse credits itself with an authority which compensates the reality from which it is exiled. That is to say, there is something that cannot be named or grasped which eludes his contemplation. The story he tells in some way narrativizes the functioning of the signifier God, but a signifier that always produces an unknown and is always outside him, and therefore always an other. However, in attempting to embody the two objects of contemplation Sufi Baba attempts to bridge the gap between the inside and the outside, the private and the public through, what de Certeau (1986) calls, an 'intimacy with exteriority'.

In embodying the two images of the virgin and the prostitute Sufi Baba conceives of himself as feminine. He corresponds to what Corbin (1969: 157–75), in his study of Ibn Arabi, calls the 'creative feminine'. In contemplating the Godhead through a concrete mental vision, the latter must not only represent the image of God, but also make the contemplator conform to this image. To do this it is necessary to undergo a second birth. Further, Corbin says a mystic obtains the highest theophanic vision (i.e., of ingesting the divine) in contemplating the image of feminine being, because it is in the image of the creative feminine that contemplation apprehends creative divinity (1969: 159). Sufi Baba, we have seen, not only undergoes a second birth, but also embodies, through an active imagination, the virgin and the prostitute. The feminine is not opposed to the masculine. Rather it is seen as combining the two aspects of the receptor and the actor, the virgin and the prostitute.

In Sufi Baba's discourse the feminine is dominant, a discourse that eventually attempts to lay bare a hidden mystic truth. In so doing, the discourse returns to the body and through that to a shadowy paradise no longer located in language. This means that Sufi Baba's body knows not what it says, and that the autobiographical text of the body laid bare is offered for the interpretation of the other. It awaits a foreign exegesis. It only becomes aware of itself in its alteration.

In this chapter I have attempted to constitute an imagination of the body which provides a means of producing signs. These signs represent (through gestures and words) items in what is considered a prior reality (given in the classification of the body) for an active human subject (Sufi Baba as one who sees and one who mediates). Sufi Baba activates his imagination through the agency of, for want of a better word, two meta-signs, given in the figures of the virgin and the prostitute. These signs initiate the imagination and affect changes of code — from the gestural to the graphic, and the author to the authored. We cannot, therefore, say that what we have is a simple picture of an independent reality of objects. Sufi Baba's imagination necessarily moves into signification in that there is a shift from object to sign (given in that through the virgin and the prostitute one can talk of divine love), and from presentation to representation (he authors signs and is authored by them). In each of the domains he participates in a prior imagination exists anterior to the process of signification, but one which both initiates the process of signification and participates within it as a constituent sign.

Chapter Eight

Conclusion

As presented in this study, the signifying limits of work, ritual and biography are framed within a combination of discourse and practice. In this frame the temporal and spatial divisions of the social structure are highlighted, the body is delimited (as the labouring body in the case of work, the domesticated body in the case of the circumcision ritual, and the feminine body in the case of the biography), and with the privileging of the other, a mode of worship, of both the community and its individual members, established. This mode of worship, it has been shown, exceeds the limits of the social structure. The frame within which the social structure, the body and the other are located is one that relates the gestural to the verbal and the technical to the social. Rather than reiterate the specific issues that emerge in each chapter I will conclude with reflecting on how the divisions of the social structure are marked on the body of individual subjects and the community. Finally, through a conception of the 'other' I will show how the social structure is transcended.

In work we find types of skills indigenous to the specific occupation. These skills and techniques constitute the world of the craftsperson in a way that it can be experienced directly and comprehensively. This work, thus, represents a coherent world, a province of meaning complete in itself, to which the worker responds in terms of an attitude of both everyday and extraordinary life. Of fundamental importance is the reiteration of an organic tradition of work where the focal point is the worker and the instruments and objects with which s/he works reflect the worker's craftsmanship. This work represents an 'imagined' world, one that is constructed both through doing and representing.

In the case of circumcision and the biography of Sufi Baba a similar parallel exists between speaking and doing. In the ritual we find that both the speech on circumcision and the gestures associated it are present on the body of the boy. These gestures (the physical act of preparing the novice and the actual operation) and words (the name designating someone and the enunciation of the azan) organize the construction of khatna. Yet, we discern a hidden inscriptional process, one by which the body enters language. In this sense, the gestures of the ritual prompt an understanding contrary to what they indicate. This understanding reaches its fruition in the everyday discourse where circumcised bodies are appendages to the discourse. The point is that the domain of the ritual as much as that of work, is ordered by a combination of saying and doing.

In Sufi Baba's therapeutics, his reading of impairment is premised on his ability to comprehend signs. He makes this possible by taking on the impairment of the other and reflecting on it according to his classification of a balanced body. In this sense his embodiment of impairment creates as many simulacra as there are persons who seek his help and to whom he responds. Yet a complete reading of this vision is possible only when he is able to feel the source of this impairment. In this sense his therapeutic gestures resonate with his function of sight. This complicity of sight with touch constitutes his memory of a cure and is imbued with a new intentionality in that for a cure to be effective his reading must be placed in the future.

To develop on the combination of discourse and practice in work, ritual and biography I will examine the verbal and non-verbal gestures entailed by each of them. As far as work is concerned, the gestures are described by the body techniques and the simultaneous verbal actions implicit in the production of yarn, cloth and quilts. Together, technique and verbal action lead to the constitution of the design of the producer, which is also a statement, in its most general terms, about the constitution and reconstitution of an idealized world. In ritual, we find that the body becomes the site for the inscription of both gesture and word. The relationship between word and gesture, in emphasizing a triadic classification of the body, establishes a balance between the corporeal body in the service of the domestic group and a spiritual body which owes its allegiance to Islam. This balance is broken in the everyday discourse on circumcision which, in denying the thematic presence of the body, establishes a self-conscious definition of the community. In the biography of Sufi Baba, discourse and practice are combined in a way that reflects on his mode of worship. This mode, combined in the two figures of the virgin and the prostitute, refers to two different practices by which he is related to the Ansari community.

CRAFTWORK

The Body Techniques of Craftwork

A dominant theme in the Ansari notion of work is the disciplining of the worker's body. This discipline can be conceived of in three ways. First, it manifests the internal rhythms of the body, and is not based on a technological determinism. These internal rhythms refer to the biography of the body in the domestic group. Here, the time of the body is punctuated and its career broken up into segments. This division of the time of the body is marked by the act of weaving and quilt making.

A second disciplinary mechanism inscribes on the worker's body objectifying techniques of work. This inscription is seen in the gendered and generational division of work. All workers involved in the weaving cycle occupy positions determined by their gender and the generation to which they belong. This position, in turn, conditions both the type of work to be performed and its execution. In this execution each worker is taught to enact certain functions which he or she must embody. Thus, for instance, we find children being instructed on the use of their feet while curing yarn. The sharpest indicator of the embodiment of functions is found when cloth for the shroud is made. Each category of worker brings to bear his or her signature to the work being done. This signature, as we have seen in the third chapter, corresponds to bodily substances: phlegm, semen and milk and teeth. The signature of each category of worker reproduces the gendered and generational positioning of workers within the social structure. It must not, however, be forgotten that in emphasizing these positions the worker's body, in marshalling the productive forces of its society, symbolizes a network of power relations. In the sense of expressing such relationships the worker's body emits signs. Because these signs (the composition of the household on the basis of gender and generation, the opening and closing of the spatial boundaries of the house, the importance of the head of the agnatic line, etc.) reside in the social structure they delimit the natural movements of the body.

A third type of disciplining of the worker's body is found in the relationship of the worker to the loom. The worker's body is divided into directions, attitudes and postures. The loom is similarly divided. Both the loom and the worker's body involve a set of correspondences and transferences that define work tasks inscribed on the body as the capillaries by which power is expressed. The second and fourth chapters document the division of the labouring body, while the third shows that in making cloth for the shroud these divisions express a notion of power anchored in the body of that community. Power, however, is not limited to the divisions of the social structure. It is also fortified by the tradition of weaving. The attempt to inscribe the corpse with carnality and spirituality by clothing it with the four elements of nature and its eventual transcription on the loom as an ancestor, legitimate work on the loom as a primordial act of creation. This act is documented in the *Mufidul Mu'minin*.

There is, however, another way in which the Ansaris relate to the work process. As weavers, the relationship of the body to the loom is legitimated through their kitab. As quilt makers, the Ansaris reflect a different conception of the body. Here, the same work material (rags and thread) has different connections, different relations of movement and rest. In this work process the body is conceived by its function, but also by its affects. The labouring body is characterized by staccato movements, and we see that a significant part of Miriam's discourse dwells on the pain visited on the body while quilts are being stitched.

The movements of the quilt maker's body articulate a space which the quilt maker calls her own. This space constitutes both the worker's body and the quilt maker's place of work. The space of quilt making is not formulated by that of weaving but is intrinsic to the stitching of quilts. It may be possible to read the working space of quilt manufacture as transgressive of that constituted by weaving. The fifth chapter makes a somewhat different point. The quilt maker expresses a fractured space of the body in that her work refers to a series of dispersed movements. The fractured space of quilt making is imbued with an affect and is, in this sense, inscribed with a multiplicity of individuating techniques. This is seen most clearly when the embroidered quilt is being fashioned — designs are not replicated but reoriented. Affect here is something that is not merely verbalized, but is strongly tactile, referring, as it does, to a sense of touch. For this reason it is possible to argue that the tactile body shapes the quilt. In shaping the quilt, the quilt maker exemplifies, what I have called, a relationship of resemblance, between the worker and the quilt and between her and other workers. Affect is a sign of sovereign power evident in that it not only shows an idea of work different from that of weaving, but is also the way the woman becomes the mistress of her conjugal hearth. Also, through quilt making women are able to exclude men and articulate an exclusive world.

Together, the body techniques of weaving and quilt making highlight a disciplinary relationship of the labouring body to the instruments of work. Discipline constitutes the labouring body by placing it both within the social structure of the community and a work tradition. This tradition is derived as much from the *Mufidul Mu'minin* as from the lal kitab. In locating the labouring body within the social structure and a work tradition, the Ansaris establish a relationship between a corporeal and metaphysical body. This is found when cloth for the shroud is fabricated: the corpse is simultaneously carnal and spiritual.

The Discourse of Craftwork

If the labouring body is fashioned in relation to the implements of work and by incorporating into the act of production symbols external to the act, it must be recognized that the moment of production is invested with a discursive authority intrinsic to the act of work. This discursive authority is evident in the conjoining of work to worship, the discursive learning of skills and the relationship of the verbal act to the movements of the labouring body.

The relationship of worship to weaving is developed in the *Mufidul Mu'minin*. The fourth chapter documents this link by showing how the loom, as material object, is classified and the various practices to which it is put. Intrinsic to the classification of the loom and the practices associated with it are a series of utterances known as du'a. Du'a, or supplicatory prayers, constitute weaving as sacred work by showing that the operation of the loom is an act legitimated by the tradition of weaving. Supplicatory prayers legitimate the work of weaving in two ways: they posit a continuity with the primordial act of creation; the authority to posit is not open to every weaver of the community, but is embodied by certain categories of men (the male head of the hearth and of the agnatic line, and the holy man). In other words, this authority operates through the utterances of the work tradition. From within this tradition this authority is distributed among those categories detailed above. The discursive mark of such authority and its distribution is found in the way each of these categories of men enunciate the prayers of the *Mufidul Mu'minin*. Thus, verbal intonation is a mark of authority.

In sharp contrast, the discourse of women as quilt makers is not built into the act of stitching the quilt since such manufacture is not premised on a formal structure of rules. The discourse of the quilt maker, when she is stitching her quilts does, however, mark her with the authority of transgression in that the boundaries established by the discourse of weaving are violated. Miriam categorically tells her husband to go back to his 'du'a salam'. The designs on the embroidered quilt transgress those represented by the world of weaving. Quilt makers often use motifs and colors which would be considered sacrilegious in weaving. The point is that significant designs of quilt making enter into the lal kitab and are remembered via verbal intonation and nonverbal practices.

Among weavers the utterances of the *Mufidul Mu'minin* are connected to the movements of the body. The fourth chapter shows that the enunciation of prayers is synchronized with the hands and feet of the weaver at work on the loom. Also, the conversation between Adam and Jabra'il assigns specific prayers to each motion of the body. In repeating these prayers the weaver invokes the loom as the memory and continuity of a work tradition. Here, the loom is a construct of language evident in that each piece of the loom is assigned a specific term and the semantic domain of each term is mapped out in a way that sounds embody virtue. The operation of the loom, in this sense, is an act of worship.

If work on the loom is an act of worship it is important to bear in mind that only specific categories of workers are authorized to enact the worship. Such authority marks the workers as those who can transmit both the loom and the skill of weaving. The third chapter mentions that a skillful weaver is one who is 'strong of speech (kalam ka pakka)', and observes that in transmitting his skill to the next generation he is referred to as 'pir sab' by the novice. In this way the loom is invested with discursive authority: the skills of weaving are a question of acquiring mastery over the utterances (du'a) and in synchronizing these utterances with the movements of the body. The legitimation of such authority is fully achieved when the worker makes weft members for the shroud. The most general attribute of weaving is characterized by the quality of the pen (qalam), descending to letters (kalimah) and then speech (kalam). These three attributes are associated with cloth, which is fashioned only after the loom has been in operation.

Circumcision and the Body

Just as the labouring body is disciplined through the techniques of work so also the circumcision ritual socializes and legitimates the body of the male novice through a series of gestures and a series of words. From the preparation of the novice a day before his operation to the removal of the prepuce we find the novice's body is in the process of being constituted and manufactured for entry to the domestic group. Simultaneously through a series of words we find the body also enters into the community of Islam. Thus, with gestural inscriptions the social structure is embossed on the body of its male members and with verbal inscriptions the body claims an Islamic heritage. Together, these two series of inscriptions highlight a triadic classification of the male body into a depth, a surface and a height. In this section I will explore the gestural inscriptions on the body.

In distinction to the everyday discourse on circumcision, the ritual situates the body as the main referential object that is carved out. This is achieved by inscribing the ritual wound, but also by socializing the body into various hygienic practices. Gestural inscriptions on the body are available from the accounts and actions of two types of actors: the mother and the barber. In their different though related ways, both argue that the novice's body is composed simultaneously of male and female characteristics and that after the ritual the body is ready for entry into the productive and reproductive life of the domestic group. To this end, the ritual legitimates the physical object of the body by inscribing on it three types of signs: ghusl, istibra and kalimah. With the first two the body is socialized into a hygienic regime and a legitimate sexuality, and with kalimah it enters into the community of Islam.

The mother's account flows not only from her position in the domestic group but also from her child's entry into it. In the case of both Miriam and Shabnam this account is tinged with nostalgia: the child's career in the domestic group is plotted in a way that will progressively distance him from his mother. For Shabnam, this distancing is expressed as her son's entry into Islam, while for Miriam the child becomes his father's son after the ritual. The barber, on the other hand, is situated outside the ritual. She is neither a member of the novice's domestic group, nor does she have an interest in his career. Her task is to initiate this career in accordance with a framed picture of an already constituted world. Yet, both Miriam and the barber believe that after the ritual the boy's body is composed of male and female characteristics. Miriam finds this combination in the relationship between milk and blood, while the barber designates this combination as 'hamdami'. The importance of the mix of male and female is that with khatna men establish a legitimized sexual community with women.

If through the combination of male and female the novice's body is prepared for conjugality, the ritual also forces the body into a hygienic regime and in this way makes its pure. As a major purificatory ritual ghusl cleanses the body of its secretions and of those elements that cause dirt and decay. In so doing ghusl attempts to exercise control over the emissions and secretions of the inner body with the intention of bringing the body to exist within the limits of socialized purity. Through ghusl, then, the novice regulates the depth of his body. Istibra, on the other hand, focuses more specifically on the genital zone by establishing an elaborate procedure for cleaning the last drop of urine. More than ghusl, istibra is especially recommended after sexual contact. Together, ghusl and istibra make the body pure by excising it of elements that impinge on its tidy insularity. This is achieved by opening the inner body to the gaze of the male and by regulating at the surface all dirt and decay. In effect, both ghusl and istibra legitimately traverse the sexuality of the body by disciplining its depth.

Circumcision and Discourse

Ghusl and istibra are modes of non-verbal practice by which the pollution inherent in the male body is brought to exist at its surface. Just as in work where the labouring body is invested with discursive authority, so also in the ritual the novice's body is written upon by the word. The removal of the prepuce and the whispering of the azan and his formal name in his right ear are synchronic events. Further, if in weaving, the movements of the body are related to utterances, in khatna this argument is carried further by showing how the body itself is simultaneously corporeal and spiritual. That is to say, in weaving the labouring body acquires a rhythm and cadence by conjoining the physical gesture with the word. This conjunction of word with gesture occurs as the self-conscious act of the weaver. In the ritual, the body itself is constituted by the physical act and the word, a constitution achieved through the agency of the other. Thus, one may argue that weaving makes manifest what is implicitly present in the ritual, and in fact completes what the ceremony begins.

In constituting the body of the male the ritual invests it with sound. Here, a deliberate attempt is made to establish a metonymic relationship between the azan and the boy's formal name so that he becomes a member of the ummah. The azan does something more than annex the boy to Islam: it facilitates the communication between the gestural and the verbal since through its recitation a link is forged between the novice's formal name and the inscription of the ritual wound. Because the azan and the formal name are whispered in his right ear we find that the boy's body is constituted as much by its corporeality (the ritual wound, the observance of a legitimate sexuality and various taboos dealing with hygiene) as by a source external to such corporeality (the azan and the name). I call this foreign source a celestial height because the boy is thought to be impregnated with the word of God. For this reason, his body is composed of a depth and a height. Together, these two are co-present on the body's surface, signifying a 'biunity'. The surface, in the sense of a biunity, is composed of a series of signs that regulate the boy's behaviour in his domestic group and the community at large.

With the discursive aspect of the ritual the body comes to exist within the elements of language: it acquires a name and individuates the liturgy. The linguistic nature of the body is pushed further in everyday discourse. Here, the significance of the ritual wound, as delimited in the ritual through gestures, recedes into the background. Eventually, gestural inscriptions on the body (specifically the value of ghusl and istibra) are absented in this discourse. In describing the contours of everyday speech we find a paradox: on the one hand a referential object is not carved out as it is in the ritual, on the other hand, to the extent that this discourse constitutes a community, we find the emergence of a master narrative, one which establishes the limits of being Muslim. This establishment is based on an imaginative identification with other male members of the community. In turn, identification is made possible through an act of memory: what is recalled in khatna is not so much one's own operation, as much as its enactment on someone else. In this recall musalmani valorizes iman and ruh.

The everyday discourse on circumcision, focuses, first, on the absenting of the body, and, second, on the expression of the community. The body, so carefully marked in the ritual of khatna, becomes in the domain of musalmani productive of a simulacra. The everyday speech of musalmani absorbs the body within itself, first by establishing a correspondence between speech and gesture and later by providing incorporeal valuations of the body. In so doing, musalmani precedes khatna and in fact engenders the latter. As evidence the representational imagery of khatna is engulfed by terms such as belief, pain, witness and so on.

The question, then, is what is being simulated? The speakers do not talk of the referentiality of the body or any of its substances. Their discourse is based on a proliferation of the signs on the body, and of seeing such signs as the epiphany of the community. These signs become a strategy by which the term Muslim is discursively mapped. There are three steps by which the simulacra is produced. First, the discourse of musalmani, in establishing a correspondence with the operation of circumcision, reflects on the body (the body as the retention of semen, as strength, as belief, as spiritual). Second, this discourse masks the body in that the significance of the inscription of the liturgy on the circumcised body exceeds gestural inscriptions on the body. In fact, the latter are prescribed only if the liturgy can be voluntarily recited. Finally, musalmani replaces the referentiality of the body by linking the community to other terms. There are two strategies by which the community coheres around musalmani. The Muslim is separated from the Hindu in a way that the Muslim's body becomes a metonym of musalmani. Secondly, the term Ansari (designating the community) enters into a duplicative relationship with terms such as iman, ruh, azan. Experientially, the sense of community is forged with pain: musalmani forces one to recognize the pain of khatna. This recognition is based not so much on one's own circumcision as on having witnessed the enactment of the ritual on the other. In this sense the community is recalled in each instance of the ritual.

Biography and the Body

Both work and the circumcision ritual supply a disciplinary matrix within which the body is situated. Within the ritual the physical object of the body is legitimated, is fabricated for entry into the domestic group and is composed of male and female characteristics. In the domain of work, the biography of the body is further punctuated, it is inscribed by those divisions of the social structure that exist outside the domestic group. In learning the details of work the body is divided into directions, attitudes and postures, as much as it acquires a tactility. Finally, a work tradition is embodied. In focusing on an alternative imagination of the body, the seventh chapter shows how, within the Ansari community, a transgressive body is constituted. This transgression is consciously enacted and embodied by Sufi Baba, an Ansari who is not a weaver, a member of the community, but one who does not belong to any of its known social divisions.

Just as the circumcision ritual composes the novice's body as spiritual and corporeal, so also Sufi Baba's body is simultaneously spiritual and sensual. His body is, however, transgressive in that he consciously elicits a recognition of the feminine body by his bodily practices. These practices are organized in two domains: the private where he cures ailing women within the confines of their domestic space, and the public where he deliberately draws attention to the feminine aspects of his practice in a space that is outside the house. He links the private and public through an embodied mode of contemplation. In this way he distinguishes between, what I call, inner voice and outer speech. Through the former his body author's signs and with outer speech his body is authored by the community. In either case his body becomes the site for the production of meaning.

In the first domain of practice — Sufi Baba's relationship with women in distress — the body is classified into an inner and spiritual zone and an outer and sensual one. Following from this taxonomy the body is composed of qalb, ruh, nafs and jism. Starting with jism, the outer and most material body, we find a progression to the most subtle body — the heart. His therapeutics is premised on uncovering the heart so that he can realise his object of contemplation, the untouched virgin. Consequently, his attempt to provide succor to women in distress has as its interior motive a mystical union with God. Accordingly, he categorizes impairment, both physical and mental, in a way that is consistent with his classification of the body. All impairment, an imbalance between the spiritual and the carnal, further subdivides the body into an upper and lower zone. The former is composed of earth and water and the latter of fire and air. This classification premises a cure on the combination of the four elements.

For reading pain Sufi Baba relies on the movement of his hand and gaze over the affected area and thus activates his classification of a balanced body. His classification is imbued with an intentionality: the movement of his hand, resonating with his experience of previous impairments, leads to an uncovering of the malady. Thus through a sense of touch the cure is as much a part of his memory as it is located in the future. In cases of 'pain of the heart' he prescribes a cure after willing a sympathetic pain in his own body.

As healer Sufi Baba authors a cure. As a mediator his body is authored during Chahullam. The various roles he plays here are organized around a night/day axis. At night Sufi Baba dissolves sexual boundaries by embodying a passive and active femininity, while during the day he acts as spokesman of the community. As an androgynous clown he invites mockery and ridicule. In the process the feminine aspects of his body are marked in a grotesque and mutilated way. It is almost as if in taking charge of his performance the community circumscribes his discourse and places it within an accepted understanding of the body. Yet, Sufi Baba does not react passively to this authority. He, too, is engaged in a politics which seeks to publicly recognize the issue of femininity.

On the other hand, as spokesman of the community he contributes to its formal definition, but in a way that his body is not engaged in the task of producing meaning.

The androgynous aspect of his body is enhanced in his role as mediator between customer and prostitute. As mediator he theatricalizes his body by showing from the prostitute's perspective how women copulate with men. In identifying with prostitutes he creates a verbal space for a recognition of their services. In turn, the customers react by mocking his body, but in a way that reflects on his masculinity. Here, his gestures as prostitute are not commented upon, yet it is as mediator that he is subject to verbal abuse that is not comic. This suggests that in enacting the gestures of prostitutes his performance threatens the masculinity of the onlookers. His role as androgynous clown and mediator forces him into a state of alienation, which he says is an exile. The ordeal of exile is expressed through words and deeds.

Biography and Discourse

Sufi Baba's biography is understood as the relationship between the intransitive and transitive. The former involves a type of practice by which he makes decisions, which in turn, are mediated through the body. The transitive feature of his biography is conditioned by factors beyond his control, though here too his body is marked by his practice. Common to both the transitive and intransitive is a contract by which he verbally interacts with the other.

There are three steps marking the discursive contract with the other. The first concerns the preconditions of discourse by which he interacts with ailing women, prostitutes and male members of his community. The preconditions are found in his mode of worship — the figures of the virgin and the prostitute — where he seeks mystical union with God. In effect, through such worship he establishes contractual relations. The second step points to the status of the contract by which he circumscribes a verbal space, but one where a large part of what he says is either held in doubt or ridiculed. With ailing women their speech is both recognized and internalized. The last step refers to his representations of establishing a story: of illness, of the geography of the female body and of the self representations of the Ansari community.

In terms of the preconditions he determines a course of action where he authors the rules of his practice so that he dialogues both with others and himself. This is achieved by establishing a relationship between his classification of a balanced body and bodily breakdown. Here, dialogue presumes the other's belief, as much as it is eventually anchored to his mode of worship. To the extent that all cure rests on a willing compliance of the other, dialogue effects a closure. Further, because mystical union with God is always elusive, he draws from his own experience to produce meaning. This meaning is not solipsistic since he pegs his experience to an object (the untouched virgin) that is external to his body. This externality is the precondition of faith. In this sense, he attempts to give his desire — the feeling of being in love — an external reference. With ailing women this is found in the untouched virgin and her perfect body. This leads to his classification of a balanced body and a therapeutics is premised on a deviation from this balance. In Chahullam the reference, found in the prostitute, proposes a carnal geography of the body. Sufi Baba's desire, in embodying the prostitute, recognizes that carnal love for the other is a manifestation of divine will. Both the virgin and prostitute create a space for recognizing femininity.

This space is not merely territorial (the private within the house, the division of public into a subjunctive and injunctive zone), but is demarcated by assigning to the contracting parties their reciprocal place: the healer and patient, performer and spectator. In both cases a topography of personal pronouns effects this distribution — with ailing women he authors a cure by embodying their pain, and in the festival he responds to questions by using the 'I' (at night) and the 'we' (during the day). As healer and clown the use of the I represents a passive and active femininity by speaking the language of the other. When he uses the we his statements tie the questioners (the you) to those who answer. The I or my is a space where the discourse of subjectivity and individuality is constructed. It is produced by its speaker. The collective we (biraderi, Mohd. Ayub, Sis Ali) points to a shared belief. Here, the we plays the role of a shifter by bolstering the tradition with reliability.

In Sufi Baba's story the I is both figurative and symbolic. It exists in the shadow of the two organizing images of the virgin and the prostitute. Thus, his discourse marks the place where the other speaks. By recourse to his mode of contemplation he speaks of femininity, indeed embodies it. Through embodiment he attempts to unite opposites: public/private, healer/patient, outside/inside, active/passive, prostitute/virgin. However, because his object of contemplation constantly eludes him the resolution of these opposites must remain alien to him. It is in this respect that he considers himself an exile. The point is that he activates his imagination by recourse to the two figures. This imagination — the best example of which is his dream — opens a space where he expresses the unity of opposites. With ailing women it appears in the schema of a healthy body and with the community this space narrativizes the I.

Work, Ritual and Biography

The constitution of the body and a discourse organized around and over it resonate in the three areas of work, ritual and biography. While there are strong similarities, and one might say, homologies, in the relation between bodily practices and discourses, both within specific areas and across them, we also find that each of these domains is concerned with specific issues. Consequently, the relation between discourse and practice as presented in the case of the Ansaris, cannot be ordained through a grammar of social life. There is, however, one common theme running across the three domains, one that concerns the yoking of temporality to the recognition of the other. In this union, the image of the community is both fabricated and reproduced.

In the circumcision ritual the community is inscribed on the body of male novices, while in manufacturing cloth and quilts this inscription is reiterated, both through bodily expression and verbal gestures. Such expressions constitute the career of individuals within the domestic group, as much as they describe the time of the domestic group and the community at large. Simultaneously, the life of individuals, domestic groups and indeed of the weaving community is framed within time: the story of origin mentioned in the *Mufidul Mu'minin* and the importance of zamana in validating the tradition of weaving are just two examples. This frame explicitly recognizes the other, both as a substantial entity and as a structure of the possible. The case of

Sufi Baba is more complex. In transgressing the orthodox divisions of Ansari social structure he does not operate or express its radical negation. Rather, his imagination of the body and a discourse centred around it can be seen as the other of Ansari social life.

Craftwork, Time and the Other

In this monograph I have argued that craftwork is organized on the basis of a division of labour. Second, the loom as material object is related in an instrumental and expressive way to the community of weavers. Third, a technical nexus between the instruments of work and the labouring body is normatively ordered through language. These three dimensions of craftwork constitute an ideal community, establishing a continuity with the work tradition of the weavers.

In the weavers' conception of work the division of labour is an agent of social cohesion in two respects: substitution and specialization. In everyday life each of the personnel involved in the four stages of weaving can be substituted by other members of the community. The second chapter shows that substitution is ordered on the basis of either kinship or contractual obligations, or by a combination of both. Thus, the specialization and interdependence of functions, given in the act of work, are found in the social structure of the community. The division of labour in terms of substitution and specialization highlights a reciprocal relation between men and women, on the one hand, and different generations on the other. By a reciprocal relation is meant that through substitution and specialization work is organized on the basis of an exchange.

However, specialization is not dictated solely by the gender or generation to which one belongs. As skill, specialization is transmitted and learned, whether this be the specialization of manufacturing cloth or quilts. By learning skills the worker acquires knowledge over the task at hand and is simultaneously socialized into the community of Ansaris.

While the division of labour is premised on a relationship of exchange, the loom is the focal point around which exchange is both organized and legitimated. This exchange is not merely between weaver and worker, within and outside the household, or the domestic domain and the larger community, but also between weavers and their work tradition. In so far as work on the loom is considered an act of worship, weaving acquires a sacred dimension. The sacred nature of weaving reinforces divisions of the social structure. Weaving is valued not merely because it is economically remunerative, but because by working on the loom the worker fulfills his obligation as weaver and as a Muslim. the sacred dimension of weaving refers to the constitution of the world given in the questions asked by the first man, Adam and the archangel, Jabra'il's replies. The constitution is an ideal one: it can never be replicated, but must, in every act of working the loom, be remembered and reiterated. Work on the loom is a means of remembrance of what the community is and the celestial abode from which it has descended. In this sense, work on the loom is a means of gaining knowledge in that it makes possible a return to the world of archetypes. Here, weaving reflects the truth of the community to the extent that it is sacred, and it emanates the presence of the sacred to the extent that it is true. In effect, work on the loom is a reminder of the beginning of time where both physical act and verbal sound are saturated with an ethical mode of work, a mode where work is inseparable from worship.

For this reason, the loom emits signs in harmony with the tradition of weaving. These signs are embodied by each individual worker and his or her ancestors. The third chapter shows that each category of worker involved in making cloth for the shroud marks work with a distinctive signature, and in so doing marks the body of the deceased. The designs represented on cloth signify the distinctive bodily traits of each category of worker. The body of the corpse is identified by its own characteristics. In this way a continuity is established with one's ancestors, one that is represented on the loom as pole (qutb). The pole is life in the above sense: it incorporates in its operation the work of this world, as well as the vision of the patron saint of the weavers, of the ancestor, of the teacher and learner.

Together, the division of personnel in weaving and quilt making, the place of the material instrument in the community of weavers and the relationship of the body to the work process show how craftwork codes time. In both weaving and quilt making the emphasis is on the manufacture, more than the career, of the product. In emphasizing the process of production we find that the worker symbolizes a range of meanings that exceed the utility of the product. The most general meaning built into the product is a work tradition. Every act of production, to the extent that it recites the prayers of the *Mufidul Mu'minin*, reproduces this tradition. In turn, this tradition, to the extent that parts of the loom and the body are delineated, constitutes the labouring body and the loom. That is to say, in manufacturing the product the artisan reiterates the story of the origin of weaving and its subsequent career in the world.

In laying out the origin of weaving this tradition, we have seen, classifies the loom into its component pieces. This classification is ordered along a light and dark axis. This contrast divides the time of weaving in filigreed detail: the day, the week and the ritual calendar are mapped through this axis. The commencement of the four stages of weaving follow this contrast. Finally, the light and dark division is spatialized in the territorial area of the hearth. Each act of producing cloth embodies the work tradition in the sense noted above.

There is, however, another way in which craftwork constitutes time. This is the time of otherness or of alteration. In this notion of time the present is a source of its imminent transcendence: whatever is contained in the present is a seed of the future pregnant with the possibility of becoming other than what simply is. Here, temporality is instituted as a socio-historical institution. In this institution craftwork is a means of making time. In turn, such time is inseparable from the time of social doing. The latter establishes the period or boundary of time. For weavers work orders the process of alteration in immediate and obvious ways. We see that each of the four stages of weaving are occupied by determinate categories of workers. The movement from one stage to another corresponds to a movement up the generational ladder. In the case of quilt makers this movement is retained. Thus, even when the Ansari community is preserving itself it does not cease to alter itself.

Through this notion of alterity the domestic hearth is perpetually reconstituted. This means that the household is not merely the location for familial relationships, but is also the ground for the emergence of the potential. The dialogical structure of weaving practices recognize the potential incarnated in the figure of the addressee. Similarly, in the manufacture of the marriage quilt the teacher/learner dyad expresses the potential to the extent that the pupil, in copying the teacher, must recast the design. To this extent the relation between past, present and future in the context of work within the household is not one of succession, correlation or disjunction, but one of repetition in that the future (through the structure of the potential) must reproduce the present, but in another form, just as the present takes the place of the past.

Just as the manufacture of the product implies an alterity so also the circumcision ritual, through an act of violence, propels the novice into an alterity. Through this alterity the boy's career is plotted in the domestic group: he will, over time, provide his labour for weaving chores, marry, raise children and become the head weaver of his hearth, and so on. In other words, this alterity is an already constituted structure of the possible by which various genealogical and affective positions are available to the boy, but in the future. In effect, the structure of the possible codes time by punctuating his career into discrete units. Thus, the past and the future are imbricated in the present.

The other is conceived of in a second way. The whispering of the azan and the boy's formal name show how the divine is ingested. In this sense, each circumcised body carries within itself the word of God and one of the names that designate him. Thus each male has a theophanic other inscribed on his body. This inscription, it may be argued, negates time since its power is of a transcendental nature. I have argued that rather than negating time, such authority initiates the biography of the social body. Further, in the context of the discourse of musalmani, the biography of the body is replaced by the self-definition of the community. This substitution is achieved by displacing on to the community the transcendental authority inscribed on bodies. The point is that in either case — of khatna and musalmani — the other is central to one's biography and the community. The other, in the sense noted above, concerns the distribution of the dimensions of time — what lies in the past and how the future is to be oriented. Associated with this is a concrete other, evident in work, which actualizes the structure of the possible in concrete situations.

In the circumcision ritual all action is inscribed on two registers of the body. The first is constituted by a sexual and material surface and the other by a spiritual one. We discern a similar operation in the biography of Sufi Baba. All possible actions are divided in two. On the one hand an entire imagination of action is projected on to the feminine body, where the action itself appears as willed and is determined in the form of restoring an ideal balance to the impaired and corporeal body. Simultaneously, this action is projected onto the metaphysical body of the untouched virgin. A second type of action, on the other hand, is neither produced nor willed by Sufi Baba, determined, as it is, by the forms of ridicule and mockery of the physical body of the woman. This action is projected on the metaphysical body of the prostitute.

These two actions or practices are structured by the other and its functioning, whether this other is concretized as the impaired body and/or the delirium of the community, or spiritualized in the figures of the virgin and the prostitute. This structure provides meaning to Sufi Baba's practice and is eventually apprehended as a way of worship. This mode of worship is embodied by him and because of this all meaning is emphasized by his body gestures. The other, then, organizes Sufi Baba's experience. Yet, this organization is not that of a closed totality since in his reading of impairment we find an active intentionality, oriented to the future, precipitating a cure. If the other authorizes Sufi Baba's practices, in that the latter are determined by his mode of worship, it is important to recognize that the other is a signifier, potentially capable of generating infinite meaning. For this reason, the other can never be fully embodied by him. It is in this latter sense that we can understand Sufi Baba's exile.

Glossary

abba	term of reference for father
abdal	substitute
achcha	okay, fine
Adam	the first man
adda	den, gathering of people
a'ge	in front, before
ajnabi	stranger
ala	instrument used in making an impression
alif	the first letter of the Persian alphabet
ama'nat	thing held in trust
ambva	the mango season
amma	term of reference for the mother
ap	term of respect or formality
apa	term of reference for sister
apne	one's own
'aql	intelligence
asbat	affirmation of God's existence
aur	and
azan	call to prayer
baba	term for an elder
bachchon	children
bade	big
badhe	to increase
badla	exchange
bahu	daughter-in-law or younger brother's wife
baithat	to sit down, a seat
baji	reference term for sister
bal	strength
balgham	phlegm
balna	wooden bar, part of the loom

ban	tree
bana	weft
bante	to make
bar	once again
bare	about
barat	third day of wedding celebrations
bari ta'ala	to remember God
bat	talk
bazari	common, of the marketplace
begair	without
beta	son
beti	daughter
bhag	run
bhai	brother
bhains	buffalo
bhanja	sister's son
Bharaichiya	Ansari group
bharna	to fill
bhatija	brother's son
bihaderi	circle of relatives living outside the kumba
bilkul	fully, absolutely
biraderi	brotherhood, the entire Ansari community
biryani	a food dish comprising of rice and meat
bobne	bobbin
bol	speak
Budh	Wednesday
buna'i	weaving
buri	bad
burqa	female garment covering the body
chacha	father's younger brother
chachi	father's brother's wife
chadana	to cover, to climb
chadar	sheet of cloth, sometimes designates cloth for the shroud
chade	to effect
chal	walk, gait
charpa'e	a bed used to seat guests
chaudai	width

chauka	one fourth, used as measure
chodne	to leave
chulah	literally fireplace or hearth, refers to a three generation dwelling
chut	slipped away
dabua	iron trough
dandiya	heddle shaft, part of the loom
dant	teeth
dard	pain, grief
dargah	a shrine, generally associated with Sufis
dastar	handkerchief
dastur muafiq	customary
davat	feast
devar	term for sister's husband's younger brother
dhancha	frame
dhat	semen
Dhuniya	cotton carder, Muslim caste group
diya	to give
dil	heart
do	two
do'ara	boundary wall
dola	palanquin
du'a	supplicatory prayers
dukh	grief
dukhyari	a grieving woman
dula'i	quilt, mattress
dulhan	bride
dum	strength
du'mat	gray clay and sand
Dushamba	Monday
ek	one
Fakir	religious ascetic, Muslim caste group
Faizabadi	Ansari group
Farrukhabadi	Ansari group
fatihah	recitation of the Quran
fikr	assign a primacy to thinking of God

gala	neck/throat
gamcha¹	shoulder cloth
gamcha²	marijuana
gara	term given to woven cloth of a particular count
garib	poor
gathuwa	bundle
gaya	left, went
gaz	unit of measure, one yard
ghagara	a long skirt worn by women
ghar	house, refers to a dwelling of two generations
gherai	depth
ghilaf	sheath, cover
ghusl	ritual bath
hai	is
haj	pilgrimage to Mecca
hajamat	hair cut
hakim	traditional doctor, signifying wisdom
halal	sacred
hamdami	of one breath
haq	right
haram	profane, sacrilege
hararat	temperature, passion
hathkarga	made by hand, manufacture
hatta	hand, grip
hatthwa	hand
hawa	greed
Hawwa	Eve
hayat	life
hidayat	righteousness
honge	it will be
hovat	is
hum	us
humse	from me
ibadat	prayer
iradah	will
ijjat	honour, prestige
ilaida	apart from

ilahi	God, divine
'ilm	knowledge
iman	belief
indi	penis
istibra	cleaning of the last drop of urine
Itwar	Sunday
Jabra'il	the angel Gabriel
jahil	non-believer, lout
jaise	similar to
jala	to burn
jamai	daughter's husband
jan	life, breath
janaza	funeral
jannat	heaven
jeth	husband's elder brother
jism	body
jisne	whoever
jodi	twin
Juma	Friday
Jumerat	Thursday
Kabariya	green-grocer, Muslim caste
kabhi	occasionally
kachcha	raw, untutored
kadha'oni	earthen pot
kafan	shroud
kahan	where
kahe	to say
kalam	speech, words
kaleja	heart, liver
kalimah	letters
kamai	earnings
kamal	miraculous
kamar	waist
kamra	room
kan	ear
kanch	hymen
kapra	cloth

karkhana	work-shed
karna	to do
Kasai	butcher, Muslim caste
katar	traditional scissors
katib	record keeper
kattari	traditional scissors
khamba	pole
Khan	Muslim caste
khana	food
khandan	extended family, used as a synonym for the kumba
khanghi	comb, part of the loom
khasi	castrated
khate	to eat
khatiya	nuptial cot
khatna	circumcision, to cut
khatra	fear, danger
khidmat	help, aid
khitan	circumcision
khun	blood
kinaya	ill-feeling
kit	where
kitab	book, text
kuch	reed, part of the loom
kuf	extended family, synonym for the kumba
kufr	blasphemy
kumba	circle of classificatory relatives usually living in a single architectural structure
lal	red
lapetan	beam, part of the loom
latif	to be pure
latifah	intense subtlety
li'l-lah	the name of God
lumbardar	supernumerary
lutf	mercy
madad	help
madarasa	seminary
Madari	Muslim caste

maike	wife's natal home
maku	iron nails, part of the loom
malum	to know
mama	mother's brother
Mangal	Tuesday
manjari	blossom
man'jha	paste used in kite flying
mannat	boon
mard	male
mardana	male section of the dwelling
mardangi	masculinity
mari	rice water
matlab	meaning
mekh	wooden nails, part of the loom
mela	fair
men	in
mian	wise man
Mirasi	musician, Muslim caste
misaj	disposition
momin	Muslim
Muhammadi	name of an Ansari group
mujawar	leader of a procession
mullah	derogatory term for a priest
murdah	corpse
murgi	hen
murid	disciple (of the pir)
musalmani	everyday term for circumcision
mutapa	fatness
nabi	prophecy
nafs	carnality
naharni	nail-cutter
nai	barber, Muslim caste
nak	nose
nakh	yarn or rags
nakhur	name of gift given to female barber
nakkara	drum
nal	shuttle, piece of the loom
nam	name

namaz	liturgical prayer
nandi	sister
naqsha	design
nashta	morning meal
nasihat	to learn or be advised on religious matters
naukari	government job
nautanki	theatrical performance
nauwan	barber's wife
nazarana	gift
neech	under, below
niyaz	secret
nur	light
nurbaf	weavers of light
nyota	invitation
pad	to study
pa'e	feet, pedals of the loom
paida	born
pakhandbaj	fraud
pakka	strong, well developed
pani	water
pankha	frame of the loom
parda	modesty, veiling
patwari	state official at the village level
pavadi	pit of the loom
peech	behind, at the back
peghambar	prophet
pehchanne	to recognize
pehli	first
peti	piece of the loom
phad	tear out
phirti	wandering
phookh	blowing of breath
phupphi	father's sister
pir	spiritual guide
potli	length of thread
Purabiya	Ansari group
qalam	pen

qalb	heart
qalbut	spool, part of loom
qayamat	judgment
qudrat	primordial nature
qurbani	sacrifice
qutb	pole, loom
rabb	lord
raftar	speed
rahat	stay
rakh	maintain, to keep
rang	colour
raz-o-niyaz	secret prayer of God
roshni	light
roti	crying
roti	wheat bread
roza	month of fasting
ruh	spirit
ruhani	spiritual
rukwai	stop
sab	everyone
sabr	patience
saf	clean
sahib	honorific title
sakat	able
sal	year
sala	wife's brother
salam alai kum	Muslim form of greeting
salat	ritual or liturgical prayer
samjha	understand
samne	in front, before
Sanechar	Saturday
saniyan	friend, beloved
sas	mother-in-law
sasur	father-in-law
sath	together
shahwar	royal yarn

Shaikh	Muslim caste
shaitan	devil
sharik	partner
shauq	passion
shuhud	vision
sir	head
sitam	tyranny
soche	to think
sui	needle
sukh	peace
sun	alone, single
suraj	sun
sut	yarn
sutli	yarn
tabbar	same as kumba, khandan
tod	break
tajarba	experience
tajjarud	free from matter
takabbur	pride
takhta	plank, part of loom
ta'lim	education
talu	palate
tana	warp members
tanzil	sending down of divine words
tarik	reclusive
tarkib	technique
tartib	sequential, orderly
tasgara	loom
tasvir	picture, image
taluk	connection
teen	three
taya	father's elder brother
taziya	a wooden and paper structure depicting a mausoleum
tuhr	pure
tum	you
ubharne	to boil over

umar	age
ummah	brotherhood
unch	up
ungal	finger
unke	his
upton	paste, oil
'urs	nuptial union
uth	stand
wahi	revelation
wahm	doubt
wajib	necessary
waqt	moment in time
wuzu	ablution
yad	remembrance
yamin	fasting
yari	friendship
yeh	this
zaban	tongue
zahir	manifest
zanana	female part of dwelling
zikr	invocation
zindagi	life

References Cited

ALAVI, H., 1972, 'Kinship in West Punjab Villages', *Contributions to Indian Sociology*, 6: 1–27.

ARENDT, H., 1958, *The Human Condition*, Chicago: Chicago University Press.

ANSARI, M.G., 1960, *Muslim Caste in Uttar Pradesh*, Lucknow: Lucknow University Press.

AYOUB, M.R., 1959, 'Parallel Cousin Marriage and Endogamy: A Study in Sociometry', *Southwestern Journal of Anthropology* 15: 266–75.

BAER, G., 1964, *Population and Society in the Arab East*, New York: Praeger.

BAKHTIN, M., 1981, *The Dialogical Imagination: Four Essays*, Austin: University of Texas Press.

BAKHTIN, M., and P.N. MEDVEDEV, 1978, *The Formal Method in Literary Scholarship*, trans. A.J. Wehrle, Baltimore: The John Hopkins University Press.

BARTH, F., 1954, 'Father's Brother's Daughter's Marriage in Kurdistan', *Southwestern Journal of Anthropology*, 10: 164–71.

BLOCH, M., 1986, *From Blessing to Violence: History and Ideology in the Circumcision Ritual of the Merina of Madagascar*, Cambridge: Cambridge University Press.

BOUHDIBA, A., 1985, *Sexuality in Islam*, trans. A. Sheridan, London: Routledge and Kegan Paul.

BOURDIEU, P., 1977, *Outline of a Theory of Practice*, trans. R. Nice, Cambridge: Cambridge University Press.

BRAVERMAN, H., 1974, *Labour and Monopoly Capital: The Degradation of Work in the Twentieth Century*, New York: Monthly Review Press.

CASTORIADIS, C., 1987, *The Imaginary Institution of Society*, trans. Kathleen Blamey, Cambridge: Polity Press.

CLARKE, C., 1970, *Beaker Pottery of Great Britain and Ireland*, Cambridge: Cambridge University Press.

COLE, D.P., 1984, 'Alliance and Descent in the Middle East and the "Problem" of Patrilateral Parallel Cousin Marriage', in A.S. Ahmed and D.M. Hart (eds), *Islam in Tribal Societies: From the Atlas to the Indus*, London: Routledge and Kegan Paul, 169–86.

CORBIN, H., 1986, *Temple and Contemplation*, trans. Phillip Sherrard, London: KPI.

——, 1969, *Creative Imagination in the Sufism of Ibn Arabi*, trans. R. Manheim, London: Routledge and Kegan Paul.

CRAPANZANO, V., 1980, *Tuhami: Portrait of a Moroccan*, Chicago: Chicago University Press.

CROOKE, W., 1974, *The Tribes and Castes of North Western India*, Delhi: Cosmo Publications, vol. III, 69–72.

DAS, V., 1973, 'The Structure of Marriage Preferences: An Account from Pakistani Fiction', *Man*, 8 (1): 30–45.

De CERTEAU, M., 1986, *Heterologies*, trans. B. Massumi, Manchester: Manchester University Press.

DILLEY, R., 1987, 'Myth and Meaning and the Tukolor Loom', *Man*, 22: 256–66.

DONNAN, H., 1988, *Marriage Among Muslims: Preference and Choice in Northern Pakistan*, Delhi: Hindustan Publishing Corporation.

EL-ZEIN, A.H., 1977, 'Beyond Ideology and Theology: The Search for an Anthropology of Islam', *Annual Review of Anthropology*, 6: 227–54.

GARDET, L., 1965, 'Du'a', in *Encyclopaedia of Islam*, vol. III, 616–17.

GEERTZ, C., 1979, 'Suq: The Bazaar Economy in Sefrou', in C. Geertz, H. Geertz, and L. Rosen (eds), *Meaning and Order in Moroccan Society: Three Essays in Cultural Analysis*, Cambridge: Cambridge University Press, 123–310.

GOODY, E., 1982, 'Introduction', in E. Goody (ed.), *From Craft to Industry: The Ethnography of Proto-Industrial Cloth Production*, Cambridge: Cambridge University Press, 1–37.

——, 1982, 'Daboya Weavers: Relations of Production Dependence and Reciprocity', in E. Goody (ed.), *From Craft to Industry: The Ethnography of Proto-Industrial Cloth Production*, Cambridge: Cambridge University Press, 50–84.

GRANQVIST, H., 1931, *Marriage Conditions in a Palestinian Village*, Commentationes Humanarum Litterarum III Helsingfors: Societas Scientiarum Fennica.

GUDEMAN, S., 1986, *Economics as Culture: Models and Metaphors of Livelihood*, London: Routledge and Kegan Paul.
HELLER, A., 1984, *Everyday Life*, trans. G.L. Campbell, London: Routledge and Kegan Paul.
——, 1981, 'Paradigm of Production: Paradigm of Work', *Dialectical Anthropology*, 6: 71–9.
HUMPHREY, C., 1971, 'Some Ideas of Saussure Applied to Buryat Magical Drawings', in E. Ardener (ed.), *Social Anthropology and Language*, London: Tavistock.
IZUTSU, T., 1964, *God and Man in the Koran: Semantics of the Koranic Weltanschuuang*, Tokyo: Keio Institute of Cultural and Linguistic Studies.
JOYCE, P., 1987, 'The Historical Meanings of Work: An Introduction', in P. Joyce (ed.), *The Historical Meanings of Work*, Cambridge: Cambridge University Press, 1–30.
KHURI, F.I., 1970, 'Parallel Cousin Marriage Reconsidered: A Middle Eastern Practice that Nullifies the Effect of Marriage on the Intensity of Family Relations', *Man*, 4: 597–618.
MARGLIN, S., 1990, 'Losing Touch: The Cultural Conditions of Worker Accommodation and Resistance', in F.A. Marglin and S.A. Marglin (eds), *Dominating Knowledge: Development, Culture and Resistance*, Oxford: Clarendon Press, 217–81.
——, 1976, 'What Do Bosses Do? The Origins and Functions of Hierarchy in Capitalist Production', repr., in A. Gorz (ed.), *The Division of Labour*, Sussex: Harvester Press.
MARSH, K., 1983, 'Weaving, Writing and Gender', *Man*, 18: 729–44.
MASSIGNON, L., 1982, *The Passions of Al Hallaj: Mystic and Martyr of Islam*, trans. H. Mason, Princeton: Princeton University Press. vol. II.
MAUSS, M., 1973, 'Techniques of the Body', *Economy and Society*, 2: 70–88.
——, 1954, *The Gift: Forms and Functions of Exchange in Archaic Societies*, trans. I. Cunnison, London: Cohen and West Limited.
MESSICK, B., 1987, 'Subordinate Discourse: Women, Weaving Gender Relations in North Africa', *American Ethnologist*, 14, 2: 210–25.
MILLER, D., 1985, *Artefacts as Categories: A Study of Ceramic Variability in Central India*, Cambridge: Cambridge University Press.

MU'MININ, I., n.d., *Mufidul Mu'minin*, trans. Murtaza Khan, Lucknow: Matba Aijaza Mohammadi.
MUNN, H., 1973, *Walibari Iconography*, Ithaca: Cornell University Press.
MURPHY, R. and L. KASDAN, 1967, 'Agnation and Endogamy: Some Further Considerations', *Southwestern Journal of Anthropology*, 21: 325–50.
———, 1959, 'The Structure of Parallel Cousin Marriage', *American Anthropologist*, 61, 1: 17–30.
PANDEY, G., 1990, *The Construction of Communalism in Colonial North India*, Delhi: Oxford University Press.
———, 1983, 'The Bigoted Julaha', *Economic and Political Weekly*, 19–28.
PATAI, R., 1967, 'The Structure of Endogamous Unilineal Descent Groups', *Southwestern Journal of Anthropology*?
PETERS, E., 1963, 'Aspects of Rank and Status Among Muslims in a Lebanese Village', in J. Pitt Rivers (ed.), *Mediterranean Countrymen*, Paris: Mouton, 159–200.
RICHARDS, I.A., 1948, *Hunger and Work in a Savage Tribe*, Glencoe: Free Press.
SAHLINS, M., 1974, *Stone Age Economics*, London: Tavistock.
SAUSSURE, F. de, 1959, *A Course in General Linguistics*, C. Bally and A. Sechehaye (eds), trans. W. Baskin, New York: Philosophical Society.
SCARRY, E., 1985, *The Body in Pain: The Making and Unmaking of the World*, New York: Oxford University Press.
SCHWIMMER, A., 1979, 'The Self and the Product: Concepts of Work in a Comparative Perspective', in S. Wallman (ed.), *Social Anthropology of Work*, London: Academic Press, 287–315.
TAUSSIG, M., 1987, *Shamanism, Colonialism and the Wild Man: A Study in Terror and Healing*, Chicago: Chicago University Press.
TRIMINGHAM, J., 1964, *Islam in East Africa*, Oxford: Clarendon Press.
UBEROI, J.P.S., 1971, 'Men, Women and Property in Afghanistan,' in S.T. Lokhandwala (ed.), *India and Contemporary Islam*, Simla: Indian Institute of Advanced Studies.
VOLOSINOV, V., 1973, *Marxism and the Philosophy of Language*, trans. Ladislav Matejka and I.R. Titunik, New York: Seminar Press.

WALLMAN, S., 1979, 'Introduction', in S. Wallman (ed.), *Social Anthropology of Work*, London: Academic Press, 1–24.

WATT, M., 1965, 'Conditions of Membership of the Islamic Community', in C.J. Bleeker (ed.), *Initiation*, 195–201.

WENSINCK, A., 1986, 'Khitan', in *The Encyclopaedia of Islam*, Leiden: Brill, vol. V (new edition), 20–2.

Index

Adam 65, 66, 116, 123, 124, 125, 126, 134, 137, 231, 246, 257
air 21, 34, 98, 107, 223, 252
Alavi, H. 49, 51
Ali, Sadiq 58, 59, 181, 185, 186, 195, 202n5, 203
Ansari, M.G. 42
Ansari(s) 1, 12, 13, 17, 18, 23, 29, 30, 31, 32, 46, 50, 52, 66, 70, 80, 125, 144n1, 163, 178, 179, 183, 184n2, n3, 185, 225, 227, 230, 234, 243, 244, 245, 255
   community 2, 16, 33, 218, 231, 251, 253, 256, 258
   conception of work 1, 10
   discourse 2
   everyday and extraordinary world 11
   kinship institutions 51
   kinship structure 65
   of Mawai 188
   practice 2
   social structure 17, 180, 256
   weavers 2
Arendt, H. 12n6
artisans 1
axis
   paradigmatic 76
   syntagmatic 76
Ayub Ansari 32, 34, 50, 66, 116, 204, 217
azan 30, 32

Baer, G. 49
Bakhtin, M. 4, 5, 8, 9, 10, 25, 33, 158
Barabanki 1, 43, 81n2, 170n3, 216, 226, 230
   Ansaris of 185n3, 218
   Muslims of 107
   weavers of 10, 217
Barth, F. 49
bihaderi 30, 37, 42, 46, 56, 66, 71, 80, 96, 99, 170
biraderi 34, 37, 49, 50, 96, 227, 254
biraderi ka khana 65
biunity 30, 180, 189, 191, 249
Bloch, M. 210
bobbins 22n11, 38, 69, 71, 84, 85, 86, 87, 88, 95, 101, 102, 104, 113, 129, 133, 135
   kunda 39, 85
body 241
   gestural 29, 30, 76
   graphic 29
   of the novice (*hamdami*) 46, 89
   verbal 30, 76
Bouhdiba, A. 184, 186, 194
Bourdieu, P. 2, 2n2, 5, 6, 7, 10, 19, 24, 30, 31, 66, 67, 118, 120, 142, 178
Braverman, H. 14
Burawoy 14n8

carnality (*nafs*)100

Castoriadis, C. 20, 35, 75, 76, 236
Chahullam 33, 34, 92, 116*n*4, 181*n1*, 215, 219, 225, 226, 230, 232, 234, 237, 252, 254
chulah (hearth) 17, 36, 37
colour (*rang*) 98
colours and designs 23, 74, 113
community 24, 27
  kinship structure 18
  Muslim artisans 1
Corbin, H. 239
corpse 21, 23, 93, 100, 101, 105, 114
Crapanzano, V. 214*n1*, 235
Crooke, W. 1

Das, V. 50
de Certeau, M. 214*n1*, 225, 229, 239
Dilley, R. 115*n3*, 127*n13*
Discourse 8–10
  and dialogue 9
  double-voiced 215
  thematic and compositional serialization 8–9
du'a 24, 27, 68, 89, 90, 105, 108, 116, 123, 126, 128, 134, 140, 245, 247

earth 21, 34, 98, 107, 223, 252
el-Zein, A.H. 179
Eve (Hawwa) 124
everyday life 11–14, 23
  production of cloth 23, 118, 129
  weaving 25
  work 11

Ferenczi 186
fire 21, 34, 98, 107, 223, 252
fission 36, 51, 62, 81

Gardet, L. 126
Geertz, C. 14*n9*
ghusl 102, 183, 185, 186, 192, 248, 249, 250
Goody, Esther 115*n1*
Granqvist, H. 49
Gudeman, S. 14*n9*

haram 51
Heller, A. 11, 12, 13*n7*, 77
heritage, Islamic 1, 16, 178
household, 17
  affinal relationship 22
  cloth production 2
  kinship and ritual organization 17
  quilt stitching 2, 4, 159, 172
  ritual 36
  space 17, 110
  women of 22*n11*
  work 36
Humphrey, C. 3*n3*, 115*n2*, 117, 117*n5*
Husain 32

Islam 32
Izutsu, T. 24, 128

Jabra'il 65, 116, 123, 124, 125, 126, 134, 137, 140, 246, 257
Joyce, P. 14*n9*
Julaha 1, 217

kanch 85
Kasdan, L. 49
khadi 1
Khan, Maulavi Murtaza 119*n8*
khatiya 39
Khuri, F.I. 49
kitab 24
kumba 30, 37, 42, 43, 46, 47, 48, 50, 56, 60, 61, 62, 63, 66, 80, 83, 86, 95, 96, 99,

104, 145, 146, 160, 161, 162, 170, 171, 173, 176, 185

labour
naukari 51
pool 44
lal kitab 28, 161, 162, 165, 167, 176, 246
Lewis, I.M. 178
loom 23, 25, 26, 56, 60, 68, 69, 70, 82, 86, 87, 88, 90, 102, 106, 115, 116, 117, 119, 120, 122, 126, 127, 131, 133, 203, 244, 257
  ancestral 81
  boundaries of work 23, 128
  community of weavers 117, 256
  everyday production of cloth 23, 117
  hereditary 61, 130
  initiation ceremony of male children into weaving 23, 118, 134
  material culture 23–6
  production for the shroud 23, 117, 120, 132
  transmission of the 23, 51, 61, 118, 137

mama 21
mardana 17, 22, 39, 63, 71, 79, 98, 99, 107, 172, 222
Marglin, S. 115$n2$
marriage 36, 51
Marx, K. 12$n6$
Massignon, L. 228$n11$
Mauss, M. 20, 75, 171
Mecca 29, 227
Medvedev, P.N. 4, 8, 9
Messick, B. 115$n2$, 119$n7$
Miller, D. 115$n3$, 117$n6$
Miriam 144, 146, 147, 150, 151, 152, 153, 154, 155, 156, 157, 158, 159, 161, 164, 165, 166, 167, 170, 171, 172, 173, 174, 175, 186, 187, 191, 192, 196, 198, 209, 244, 245, 248
moment (*waqt*) 104, 133, 135
Momin 124, 125
mother's brother (MB) 22, 43
  and the bride 22
Mufidul Mu'minin8 7, 9, 24, 26, 68, 89$n5$, 96, 105, 116, 117, 118, 119, 123, 126, 129, 137, 138, 141, 159, 217$n4$, 218, 229, 244, 245, 246, 255, 258
Muminin, Irshadul 119$n8$
Munn, H. 3$n3$, 117
Murphy, R. 49
mystic 214

Nabi, Haji Ghulam 123, 124, 125, 137
non-verbal
  action 3
  gestures 2
  practice 4
nuptial union (*'urs*) 99, 163, 193
nurbaf (weaving of light) 24, 45$n1$, 68, 120, 136, 206, 207, 208

objective structures 5

Pandey, G. 1, 68
Patai, R. 49
Peters, E. 49
phlegm (*balgham*) 92, 102, 197
pole (*qutb*) 22, 106, 114
polyphonic dialogue 26, 28
practical taxonomies 6, 24, 118, 125, 126
  and dialogue 23
  and habitus 6, 7

and objective structures 7
practice 5–8
and strategies 6
prostitute 33, 35, 225, 227, 231, 232, 233, 234, 235, 238, 239, 240, 253, 254, 260

qayamat (resurrection and judgment) 93, 108
quilt(s) 145
  embroidered 27, 145, 162, 163, 164, 167, 170, 175, 176, 244
  making 26–9, 40, 43, 144, 164
    body techniques 151
    colour 154
    design 155
    dialogue 156
    extraordinary 27
    everyday 27, 144
    measurement 153
  patchwork 27, 146, 148, 151, 160, 161, 176, 177
  room 39, 41, 71, 107, 159, 163, 183
  stitching 163, 176
Quran 179, 202, 203, 206, 208

reeling 38, 71, 80, 84, 95, 99, 102
Richards, I.A. 15
ritual(s)
  and biography 29
  calendar 22
  circumcision 2, 4, 7, 10–11, 15, 19, 43, 178, 183
    khatna 29, 31, 178, 182, 183, 187, 198, 199, 202, 203, 208, 212, 242, 250, 259
    musalmani 29, 31, 32, 178, 182, 183, 189, 198, 199, 202, 203, 208, 210, 212, 250, 251, 259
  weaving cloth for the shroud 2, 19, 20, 21, 30, 36, 45, 66, 72, 74, 79, 93, 113, 141

Sahlins, M. 172
Saussure, F. de 3, 23, 25, 117
Scarry, E. 212
Schwimmer, A. 15
semen (*dhat*) 92, 100, 202
Shabnam 185, 186, 209, 248
signature(s) 21, 22, 98, 99, 102, 113, 180, 243
  bride 21
Sis Ali Salam 32, 50, 106, 116, 119, 204, 217, 231, 254
social structure 7, 15, 17–20
spirit (*ruh*) 106
stranger (*ajnabi* 93, 108
Sufi Baba 4, 7, 10, 11, 13, 16, 29, 33, 34, 35, 214–25, 227, 229–40, 241, 251, 252, 253, 254, 256, 260
  intransitive body 219, 236
  private body 216
  public body 216
  therapeutics 222, 225, 237, 242, 252, 254
  transitive body 226, 227

Taussig, M. 214*n1*
terminology 36
tradition (*zamana*) 104
transmission ceremony 25
Trimingham, J. 178

Uberoi, J.P.S. 42
Umar, Muhammad 123, 124, 125, 144, 149, 150, 157,

159, 168, 170, 186, 198, 203, 204, 206, 208

virgin 33
Volosinov, V. 5, 9, 10, 139, 140, 141

Wallman, S. 14*n8*
warp beam 69, 79, 86, 101, 102, 135
warping 22*n11*, 80
water 21, 34, 98, 107, 223, 252
Watt, M. 178
weavers 91, 108
   cosmology 74, 113
   Muslim handloom 1
weaving, 77, 110
   body techniques 13, 19, 152
   everyday world of 20
   space 74
   techniques of 20
   time 74, 78
   tradition of 20–23
Wensinck, A. 185*n3*
work 2, 14–16, 17, 19, 26, 80, 241
   body at 20, 25, 243
   concept of 14
   conceptual model 15
   instruments of 4, 74, 113
   material culture of 13
   practical action 15
   and ritual 4
   and its temporal rhythms 23
   women's 19
work shed 17, 22*n11*, 24, 39, 40, 69, 72, 102, 104, 107, 135, 136

yarn, 72, 79, 81, 81*n2*, 82, 84, 87, 88, 95, 99, 100, 113
   curing 38, 81, 129, 130
   sizing 38, 63, 97

zanana 17, 22, 22*n11*, 39, 56, 62, 63, 72, 79, 83, 85, 86, 95, 96, 97, 98, 99, 101, 102, 107, 160, 163, 172, 173, 183, 186, 222
zikr 77, 101, 102, 108, 221